SOCIETY FOR THE STUDY OF HUMAN BIOLOGY

SYMPOSIUM SERIES : 29

The physiology of human growth

T0282586

PUBLISHED SYMPOSIA OF THE
SOCIETY FOR THE STUDY OF HUMAN BIOLOGY

Numbers 1–9 were published by Pergamon Press, Headington Hill Hall, Headington, Oxford OX3 0BY. Numbers 10–24 were published by Taylor & Francis Ltd., 10–14 Macklin Street, London WC2B 5NF. Further details and prices of back-list numbers are available from the Secretary of the Society for the Study of Human Biology.

The Physiology of Human Growth

Edited by

J.M. TANNER
M.A. PREECE

*Professors, Department of Growth and Development,
Institute of Child Health, University of London*

CAMBRIDGE UNIVERSITY PRESS
Cambridge
London   New York   Port Chester
Melbourne   Sydney

CAMBRIDGE UNIVERSITY PRESS
Cambridge, New York, Melbourne, Madrid, Cape Town, Singapore, São Paulo, Delhi

Cambridge University Press
The Edinburgh Building, Cambridge CB2 8RU, UK

Published in the United States of America by Cambridge University Press, New York

www.cambridge.org
Information on this title: www.cambridge.org/9780521344104

First published 1989
This digitally printed version 2008

A catalogue record for this publication is available from the British Library

Library of Congress Cataloguing in Publication data
The Physiology of human growth / edited by J.M. Tanner, M.A. Preece.
    p.   cm. -- (Society for the Study of Human Biology symposium series ; 29)
    Proceedings of an Annual Symposium of the Society for the Study of Human
Biology, held at the Dept. of Biological Anthropology at Oxford in Apr. 1987.
    Includes index.
    ISBN 0–521–34410–7
    1. Human growth – Congresses.   I. Tanner, J.M. (James Mourilyan)
II. Preece, M.A. III. Society for the Study of Human Biology.
Symposium (29th : 1987 : Oxford University) IV. Series.
    [DNLM: 1. Growth--congresses. W1 S0861 v. 29 / WS 103 P758 1987]
QP84.P545 1988
612' .6–dc19
DNLM/DLC
for Library of Congress   87–35186
                                                                    CIP

ISBN 978-0-521-34410-4 hardback
ISBN 978-0-521-08915-9 paperback

CONTENTS

CONTRIBUTORS

E.A. Ahmed
MRC Dunn Nutrition Unit, Downham's Lane, Milton Road, Cambridge CB4 1XJ, UK.

R.M. Blizzard
Department of Pharmacology, University of Virginia School of Medicine, Charlottesville, VA, USA.

W. Bovenberg
Laboratory for Physiological Chemistry, State University of Utrecht, 3521 GG Utrecht, The Netherlands.

C.G.D. Brook
Department of Medicine, The Middlesex Hospital Medical School and University College, London, UK.

W. H. Busby
Departments of Medicine and Pediatrics, University of North Carolina School of Medicine, Chapel Hill, NC 27514, USA.

D.R. Clemmons
Departments of Medicine and Pediatrics, University of North Carolina School of Medicine, Chapel Hill, NC 27514, USA.

P.S.W. Davies
Department of Growth and Development, Institute of Child Health, University of London, 30 Guilford Street, London WC1N 1EH, UK.

V. French
Department of Zoology, Kings Buildings, West Mains Road, Edinburgh EH9 3JT.

D.J. Hill
Department of Paediatrics, University of Sheffield, Clinical Sciences Centre, Northern General Hospital, Sheffield S5 7AU, UK.

M. Jansen
Department of Pediatrics, State University of Utrecht, 3521 GG Utrecht, The Netherlands.

Georgeanna Klingensmith
Department of Pediatrics, Children's Hospital, Denver, CO, USA.

H.D. Mosier
Department of Pediatrics, University of California, Irvine, CA 92717, USA.

J. Müller
Laboratory of Reproductive Biology, University Department of Obstetrics and Gynaecology Y, Rigshospitalet, 9 Blegdamsvej, DK–2100 Copenhagen 0, Denmark.

Jennifer Najjar
Department of Pediatrics, Vanderbilt University Medical School, Nashville, TN, USA.

A.A. Paul
MRC Dunn Nutrition Unit, Downham's Lane, Milton Road, Cambridge CB4 1XJ, UK.

P. de Pagter-Holthuizen
Laboratory for Physiological Chemistry, State University of Utrecht, 3521 GG Utrecht, The Netherlands.

M.A. Preece
Department of Growth and Development, Institute of Child Health, University of London, 30 Guilford Street, London WC1N 1EH, UK.

S. Reichlin
Endocrine Division, New England Medical Center Hospitals, Tufts University, Boston, MA, USA.

Jean Rivier
Clayton Foundation Laboratories for Peptide Biology, Salk Institute, San Diego, CA, USA.

Alan D. Rogol
Department of Pediatrics, University of Virginia School of Medicine, Charlottesville, VA, USA.

Frank H. Ruddle
Department of Biology and Human Genetics, Yale University, New Haven, CT 06511, USA.

Vicki R. Sara
Karolinska Institute's Department of Psychiatry, St. Göran's Hospital, Box 12500, S-112 81 Stockholm, Sweden.

P.J.J. Sauer
Department of Neonatology, University Hospital, Rotterdam, The Netherlands.

N.E. Skakkebæk
Laboratory of Reproductive Biology, University Department of Obstetrics and Gynaecology Y, Rigshospitalet, 9 Blegdamsvej, DK 2100 Copenhagen 0 and Department of Paediatrics, Hvidovre Hospital, Copenhagen, Denmark.

Patricia Smith
Department of Medicine, The Middlesex Hospital Medical School and University College, London, UK.

Michael H.L. Snow
MRC Mammalian Development Unit, Wolfson House University College London, 4 Stephenson Way, London NW1 2HE, UK.

R. Stanhope
Department of Growth and Development, Institute of Child Health, University of London, 30 Guilford Street, London WC1N 1EH, UK.

J.S. Sussenbach
Laboratory for Physiological Chemistry, State University of Utrecht, 3521 GG Utrecht, The Netherlands.

C. Thøger Nielsen
Department of Paediatrics, Hvidovre Hospital, Copenhagen, Denmark.

M.O. Thorner
Department of Internal Medicine, University of Virginia School of Medicine, Charlottesville, VA, USA.

Louis E. Underwood
Departments of Medicine and Pediatrics, University of North Carolina School of Medicine, Chapel Hill, NC 27514, USA.

W. Vale
Clayton Foundation Laboratories for Peptide Biology, Salk Institute, San Diego, CA, USA.

Mary Lee Vance
Department of Pediatrics, Children's Hospital, Denver, CO, USA.

J.L. Van den Brande
Department of Pediatrics, State University of Utrecht, 3521 GG Utrecht, The Netherlands.

R.G. Whitehead
MRC Dunn Nutrition Unit, Downham's Lane, Milton Road, Cambridge CB4 1XJ, UK.

PREFACE

The last time the Society for the Study of Human Biology devoted one of its annual symposia to Human Growth was in 1960, and the subject has undergone an enormous expansion since then. Much of this expansion has been in the direction of experimental and clinical physiology and this new symposium, held at the Department of Biological Anthropology at Oxford in April 1987, concentrated on these areas. We have been unable, of course, to do more than sample the work going on in each of the fields we chose: experimental auxology, infant nutrition and body composition, the somatomedins or IGFs, and the endocrine control of human puberty. But all the papers presented here are at the cutting edge of their subjects; all are by authors engaged day-to-day in the struggle to extend our knowledge and understanding.

The symposium was attended by some 50 persons from half a dozen countries, and we are only sorry we cannot somehow convey here the liveliness of their discussion and the good fellowship of their meeting. We thank them all for their participation, and we wish to thank also the Royal Society, Eli Lilly Company Ltd, and Milupa Ltd whose financial support made the meeting possible.

J.M. Tanner
M.A. Preece

PART I

Experimental studies of growth

MICHAEL H.L. SNOW

Embryonic growth and the experimental manipulation of fetal size

Introduction

The weight of a newborn mammal, which in normal circumstances is characteristic of the species under consideration, is not simply a reflection of gestation length and/or nutrition but principally is determined genetically through growth rate. Analysis of fetal weight and conception age suggests that mammals can be segregated into three or four groups each having a different growth rate. Irrespective of whether the species displays fast, intermediate or slow growth the overall rate through late embryonic and fetal development is quite smooth, with no noticeable periods of fast or slow growth. There is a gradual decline in the rate from fast in early development to a slower rate as birth approaches (Snow, 1986 for review). Significant departures from the predicted growth curve, either above or below the norm, are regarded as pathological and have a correlation with abnormalities of various sorts (Neligan *et al.*, 1976; Spiers, 1982; Gould, 1986; Jones, Peters & Bagnall, 1986).

What then is known of the controls over embryonic growth?

Maternal influences vs embryonic genotype

It is clear that maternal size is to some extent reflected in fetal size, small mothers tending to have small babies and vice versa. There is also a maternal effect associated with parity, and in polytocous animals an effect of litter size. Part of these phenomena will be a function of the genotype of the fetus and part the physiology of the mother – fetal crowding or undernutrition obviously serving to restrict fetal growth, whether the major controlling factors reside in fetal genotype or in maternal physiology. Analyses of the uterine effect on birthweight of hybrids or of embryos surgically transferred between large and small strains of various species have been recently reviewed (Snow, 1986). It seems clear that a large uterus will to some extent facilitate the growth of a small genotype embryo but a small uterus severely constrains the growth of a large genotype embryo. Most of these studies measured birth weight and therefore do not give any information about the shape of the fetal growth curve, which would show at what time the effect on fetal growth was operative. Aitken, Bowman & Gould (1977) analysed fetal weight at 16.5 days *post coitum* (dpc) in the mouse, which has a gestation of about 19 days. They found no evidence for a uterine effect at that stage, either on small genotype or large genotype embryos when transferred to the intermediate sized surrogate mothers of the parental stocks. Since in the mouse a negative correlation between litter size and fetal weight has been found at 17.5 dpc and 18.5 dpc and at birth (Healy, McLaren & Michie, 1960; McCarthy, 1965; McLaren, 1965) it would appear that the constraining influence on fetal growth is a phenomenon of late pregnancy. A similar conclusion can be drawn from the growth curves of human twin fetuses which show a lowered birth weight in comparison to singletons. Twin fetuses, irrespective of zygosity, are of comparable size to singletons until the last 10–12 weeks of pregnancy during which their individual growth rate falls (McKeown & Record, 1952; Kloosterman, 1970).

Classically, fetuses were regarded as being in competition with one another for a limited supply of nutrients, this providing a simple explanation for the observed decline in weight with increased fetal numbers. This view was challenged by the data of Healy *et al.* (1960) and McLaren (1965) who observed that there were effects on mouse fetal weight associated with the number of implants per uterine horn (as opposed to total litter size), and the position within the norm. Moreover dead fetuses (presumably no longer competing for food) exerted the same effect as live ones. These authors proposed that the fetal weight data in the mouse could be explained by relation to blood supply, with those implants nearest the point of inflow of blood to the uterus being at a growth advantage over those further downstream. However in the rat, which has a similar uterine blood circulation, the distribution of fetal weights does not accord with this haemodynamic theory. Barr, Jensh & Brent (1970), Bruce & Norman (1975) and Barr & Brent (1970) have shown that an intact arterial blood supply is not necessary to establish the characteristic distribution of fetal weights in this animal. Direct measurement of blood flow to fetuses, whilst revealing variability, does not show a correlation with either litter size or fetal position (Buelke–Sam, Holson & Nelson, 1982). Thus fetal competition, limiting nutrition and blood flow seem inadequate explanations for the fetal weight distribution in rodents. It remains plausible that the late decline in fetal growth reflects physical constraint and that the effect of position within the uterine horn may be associated with slight spatial differences in the ease with which implantation is achieved, thus conferring a few hours' advantage to fetuses at certain sites.

There is no appropriate published data of litter size/fetal position/fetal weight relationships for stages prior to 17.5 dpc in the mouse but unpublished data gathered in my laboratory on some 40 normal 14.5 dpc litters show a slight positive correlation between litter size and fetal weight. This positive correlation is increased if litter size is experimentally reduced by unilateral ovariectomy or fallopian tube removal: fetal weight is increased in such conditions (Gregg, 1985). This observation is curious and unexpected; should the positive correlation stand up to further analysis questions are raised about possible feedback mechanisms linking embryo number with a growth stimulating function in the pregnant female. In any case it seems unlikely that further data would establish a significant *negative* correlation between litter size and 14.5 dpc fetal weight, so it seems reasonable to conclude that embryonic and early fetal growth is not dictated by maternal environment, but is a reflection of embryonic genotype.

Growth control in early embryos: sex differences

Experimental manipulation carried out on preimplantation embryos which either reduces embryo size (by blastomere removal) or increases it (by aggregation together of several embryos) does not result in smaller or larger fetuses. A change in cell proliferation rate shortly after implantation compensates by regulating size either up or down (see Snow, 1986 for references and discussion). Although the upward regulation needs a novel acceleration in cell proliferation in the embryo it is not possible to exclude an influence of growth factors from the pregnant female. Similarly the apparent regulation of embryo size to a postulated 'norm' during primitive streak stages (Snow, 1986) could also be controlled by maternally derived factors. However recent data have shown sex associated differences in embryonic growth and development rate which must be inherent in the embryo. Seller & Perkin–Cole (1987) have shown that in 8.25 dpc mouse embryos the least well developed embryos tend to be female rather than male, and lag behind by about 2 somites; according to the somite number/age curves of Tam (1981) this represents some 2–3h delay. A previous study demonstrated that XO monosomic embryos lag behind XX sibs (Burgoyne, Tam & Evans, 1983) and the unpublished data on XY embryos gathered in those studies show the male embryos to be further advanced than XX females, by about 1–2 somites at 9.5 dpc. My own unpublished data show the same degree of developmental advance in male embryos at 8.5 dpc. The sex difference seems to be generated during cleavage, since transfer of early cavitating blastocysts to foster mothers yields litters in which the sex ratio is skewed towards males, whilst the late cavitating group produces more females (Tsunoda, Tokunaga & Sugie, 1985).

Genetic selection for size

In mice selected for large and small body size at 6 weeks of age a twofold difference in weight can be established (Falconer, 1955, 1973). Comparisons of embryonic development in these mice reveals no difference in cleavage rate prior to implantation (Bowman & McLaren, 1970) but it sems that growth rates begin to diverge shortly after that since a significant difference in embryonic sizes reported at around 8 dpc, at the onset of organogenesis (Gauld, 1980).

Analyses aimed at identifying the processes underlying the growth rate differences show that in the large strain mouse both cell number and cell size are increased. Cell number increase has the greater overall impact (Falconer, Gauld & Roberts, 1978) but there are organ-specific variations. In liver and kidney the relative contribution of increased cell number and increased cell size seem about equal but in spleen and lung about 70% of the increase in size was found to be due to increased cell number. It was noted that males generally had larger organs than females, significantly so in lungs and kidney. Falconer *et al.* (1978) noticed in a comparison of mice at 3 weeks and 15 weeks that the cell size increment between these times was proportional to the respective amounts of growth made in the selected lines. Comparison of the selected lines at the same body weight but different ages indicated organs of similar cell number and cell size. Falconer therefore postulated that the selection process acted on differences in the relationship between chronological age and developmental age.

It is thought that the curious double peaked postnatal growth velocity curve found in primates, including man, may similarly reflect the slowing and expansion of the developmental timetable (Watts, 1986). Other mammals show a uniform gradual deceleration in postnatal growth. In the context of the size-selected mice it might be anticipated that the larger line would show certain developmental events occurring later in the growth curve. Data of precisely this sort is not available but Blakely (1979) found that the maximum elongation rate of fetal tibias occurred one day later in *small* strain animals. If this aspect of growth rate is related to the developmental timetable of hind-limb development then the relationship seems to be the wrong way round.

Blakely's data also show that large-genotype tibias have a significantly higher growth rate *in vitro* and furthermore that culture medium 'conditioned' by addition of 'large' embryo extract consistently (but statistically not significantly) supports better growth of tibias of all genotypes. The implication that some humoral growth factor may be involved needs to be tempered by the results of a detailed search which failed to find a growth controlling organ/centre which might have been the source of such a factor (Falconer, Gauld, Roberts & Williams, 1981; Snow, 1986 for discussion).

Embryonic growth control: the whole vs the parts

Even a cursory glance over the tables of embryonic development for any species, mammalian or not, vertebrate or not, cannot fail to notice that development is normally a highly coordinated process with all parts of the embryo undergoing their morphogenesis in harmony with one another. A large volume of literature on induction and cellular and tissue interactions points to the fact that conversation between parts is essential to maintain normal development and may be closely restricted in time and space (Lehtonen & Saxen, 1986). It is equally clear that the relative proportions of organs within the animal change during the course of embryonic, fetal and postnatal development, eventually assuming the adult format. Thus parts of the embryo/fetus show their own individual growth profiles. Some of these data and the mechanisms which may be involved have been recently reviewed (Goss, 1986; Brasel & Gruen, 1986).

It has been found in the mouse that the growth of parts and its coordination can be profoundly disturbed in early embryonic (primitive-streak) stages by insult with cytotoxic agents (Snow & Tam, 1979; Snow, Tam & McLaren, 1981; Snow, 1983, 1987). Implementation of novel, changed growth profiles of parts for 3 to 5 days during organogenesis restores almost complete anatomical normality before birth and permits further development to adulthood in about 40% of the offspring. Detailed study of the embryonic development of these mice reveals some features of relevance to growth control and its relationship to morphogenesis (Tam, 1981; Tam & Snow, 1981; Gregg, 1985; Snow & Gregg, 1986).

First of all, in the general increase in cell proliferation rate that occurs throughout the embryo following reduction in cell number, ectodermal tissues seem able to raise their mitotic rate to a greater level than mesodermal tissues. Mesoderm however maintains an elevated rate for longer. This results initially in a large divergence from the normal ratio of ectodermal to mesodermal tissue, which is manifest as a neural tube adopting a wavy form along the axis of the embryo. The discrepancy in size is corrected as the growth rate in the neural tube slows while that in the mesoderm continues at an elevated level. These two cell lineages also behave differently in the morphogenetic timetable. The neural tube, after a short delay in the raising of headfolds, proceeds through development according to the normal chronological time scale whereas various parts of the mesodermal lineage (somites, limbs, vascular system, haemopoietic system) show developmental delay of different magnitudes (Snow, 1987).

Primordial germ cells suffer a population depletion in proportion to the rest of the embryo and the subsequent acceleration of cell proliferation they undergo is restricted both in duration and in the site within the embryo in which it occurs. Thus the mitotic rate is doubled only for the 24h period during which the germ cells are migrating across the mesentery from the hind gut to the dorsally situated genital ridge. This time limitation for restoration of germ cell numbers leads to gonads forming with a reduced germ cell population (Tam & Snow, 1981). As a result adult females who suffered cytotoxic insult as embryos have small ovaries with reduced numbers of oocytes and males present testes which may contain seminiferous tubules devoid of germ cells. Fertility is reduced in consequence.

The features of this restorative growth are consistent with the view that mesoderm morphogenesis requires that a critical mass of tissue is produced before differentiation can occur whereas no such constraint operates for ectodermal tissues. It cannot be claimed that tissue mass is a trigger for differentiation, since in untreated embryos structures are much larger than in treated embryos at the time when such processes commence. Thus it seems unlikely that growth/differentiation is controlled simply by titration of humoral factors against tissue mass; rather, individual cell lineages and organ systems have the ability to regulate their growth either in response to tissue-specific growth factors or by autonomous active deployment of receptors for non-specific growth factors.

Examination of the skeletal phenotype of newborn mice following cytotoxic damage at primitive streak stage reveals a very high proportion of mice with extra vertebrae in the presacral spinal column (Gregg & Snow, 1983). Analysis of the course of development of these phenotypes (Gregg, 1985, and in preparation; Snow & Gregg, 1986) shows that at the time the skeletal pattern is changed the cells which will be involved in the generation of the extra vertebra(e) are located in the primitive streak or tail bud of the embryo and will not emerge as a segmented somite until some 12–15h later. Comparison with many other circumstances which generate extra vertebrae identifies elevated growth rate as the only clear similarity between them and leads us tentatively to suggest that this aspect of skeletal patterning may be a function of cell proliferation rates at crucial stages of development.

Overall, our experiments with cytotoxic agents and the resulting changed embryogenesis provide a dramatic example of how chronological age and developmental age can be altered and illustrate some of the morphological consequences of such a shift. It is clear that although anatomical changes are found they are not of a grossly abnormal nature. Whilst it must be concluded that the high postnatal mortality is a result of the damage to the early embryos, presumably indicative of more subtle errors in anatomy, physiology or biochemistry, the existence of long term survivors who appear quite normal is testimony to the remarkable ability for development to tolerate and then correct, by modulation of organ growth, profound disturbances to its normal pattern.

Growth factors

Growth control in the embryo is clearly manifest, at both a general and local level, from stages prior to organogenesis. For these early stages there is little direct evidence for the production and functioning of either growth factors or their receptors. However, circumstantial evidence, for instance from studies with teratocarcinoma stem cells and from work on oncogenes with known growth regulating activity (see Adamson, 1987 for review), implicates several such factors.

Growth hormone does not normally cross the placenta in significant quantities and endogenous sources become active only in comparatively late fetal stages. Even in transgenic mice carrying additional rat or human growth factor genes the effects of production of extra growth hormone are seen only in postnatal mice, embryonic and fetal development apparently being normal (Palmiter et al., 1982, 1983). Thus it can be concluded that growth hormone is not active in embryonic growth control.

Insulin likewise is only produced in fetal stages (Hill, 1976, and this volume) and it also does not cross the placenta so seems unlikely to be involved in embryonic growth control. However Sadler (1980) reports a growth promoting effect of insulin on 8 dpc mouse embryos cultured in vitro and Heath, Bell & Rees (1981) demonstrate the appearance of insulin receptors during differentiation of embryonal carcinoma (EC) cells. These cells are commonly regarded as equivalent to the undifferentiated primitive ectoderm of primitive streak stage embryos, from which they have often been derived. Since insulin is teratogenic when administered via the pregnant female to embryonic stages as early as 8 dpc (Cole & Trasler, 1980) the possibility that it is indeed active in growth control in the embryo remains to be definitely resolved.

The somatomedins (insulin-like growth factors) also seem unlikely candidates for embryonic growth control since their function appears to be growth hormone dependent (see Clemmons, this volume). Nevertheless, the recent demonstration (Han, D'Ercole & Lund, 1987) that they are produced by the mesenchymal (connective tissue?) element of a wide range of human fetal organs must generate speculation that they may have a role in the control of embryonic development, particularly as mesenchymal connective tissue would be a good candidate for the growth controlling organ sought in vain by Falconer *et al.* (1981) in the chimaeric large ↔ small mice (see Snow, 1986 for discussion).

Epidermal growth factor (EGF) receptors are present in 10 dpc mouse embryos (Hortsch *et al.*, 1983) and on trophoblast tissue developed in blastocyst outgrowth, which are equivalent to about 7 or 8 dpc (Adamson & Meek, 1984). However the embryo/fetus does not appear to make EGF itself until around birth (Popliker *et al.*, 1987).

Transforming growth factor α (TGFα), which has homology with EGF and binds to EGF receptors, producing the same effects, has been found in 7 dpc mouse embryos (Twardzik, 1985), and could plausibly act via EGF receptors to modulate early embryonic growth. The other family of transforming factors, the TGFβ group, has known growth regulating functions, both inhibition and stimulation, in embryonic/fetal systems (see Massagué, 1987 for review). TGFβ is widely distributed in normal tissues and will stimulate expression of the c–sis oncogene in mouse embryo fibroblasts. The c–sis gene product is homologous to the B chain of platelet derived growth factor (PDGF) and is found to be expressed in the 4–5 week old human placenta (Goustin *et al.*, 1985). EC cells also make PDGF and it has been suggested that embryonic/fetal growth could be modulated locally by interaction between TGFβ and PDGF (Adamson, 1987).

Two other oncogenes, c–fos and c–myc, are of interest in the context of embryonic growth. The products of both genes are DNA binding proteins the c–fos protein controls passage of cells from G0 to G1 in the cell cycle and the c–myc protein regulates DNA synthesis. C–fos seems to be involved in cell differentiation and c–myc in cell proliferation (see Adamson, 1987 for review). All stages of mouse embryos that have been examined express c–myc; c–fos is expressed in 7 dpc mouse conceptuses but predominantly in the extraembryonic components. In mouse fibroblasts in which c–sis is induced by TGFβ the release of the PDGF-like factor seems to stimulate activation of c–fos and then of c–myc. Both oncogenes are expressed in EC cells, c–fos at low levels in undifferentiated cells but increasing as differentiation proceeds; c–myc is maximally expressed during proliferation of the stem cells and declines with differention. c–myc and c–sis are co-expressed in the human placenta (Goustin *et al.*, 1985). Although the relationships between c–sis, c–fos and c–myc need clarification, a growing volume of literature seems to support the notion of a cascade involving all three in that order, possibly triggered initially by TGFβ (Adamson, 1987). Unfortunately, in the context of embryonic growth where prolonged cell proliferation precedes differentiation the expression of c–fos and c–myc seems to be in the reverse order to what would be expected.

Clearly many questions concerning the deployment of growth factors and oncogene products in early embryos are still to be answered but it remains most likely that the factors controlling growth will be found amongst these molecules.

References

Adamson, E.D. (1987). Oncogenes in development. *Development*, **99**, 449–71.

Adamson, E.D. & Meek, J. (1984). The ontogeny of epidermal growth factor receptors during mouse development. *Developmental Biology*, **103**, 62–70.

Aitken, R.J., Bowman, P. & Gauld, I. (1977). The effect of synchronous and asynchronous egg transfer on foetal weight in mice selected for lafge and small body size. *Journal of Embryology and Experimental Morphology*, **37**, 59–64.

Barr, M. & Brent, R.L. (1970). The relation of the uterine vasculature to fetal growth and the intrauterine position effects in rats. *Teratology*, **3**, 251–60.

Barr, M., Jensh, R.P. & Brent, R.L. (1970). Prenatal growth in the albino rat: effects of number intrauterine position and resorption. *American Journal of Anatomy*, **128**, 413–28

Blakely, A. (1979). Embryonic bone growth in lines of mice selected for large and small body size. *Genetical Research, Cambridge*, **34**, 77–85.

Bowman, P. & McLaren, A. (1970). Cell number in early embryos from strains of mice selected for large and small body size. *Genetical Research, Cambridge*, **15**, 261–3.

Brasel, J.A. & Gruen, R.K. (1986). Cellular growth: Brain, heart, lung, liver and skeletal muscle. In: *Human Growth*, ed. F. Falkner & J.M. Tanner, 2nd edition, Vol. 1, pp. 53–65. Plenum Press, New York.

Bruce, N.W. & Norman, N. (1975). Influence of sexual dimorphism on foetal and placental weights in the rat. *Nature*, **257**, 62–3.

Buelke–Sam, J., Holson, J.F. & Nelson, C.J. (1982). Blood flow during pregnancy in the rat: I Dynamics of and litter variability in uterine flow. *Teratology*, **26**, 279–88.

Burgoyne, P.S., Tam, P.P.L. & Evans, E.P. (1983). Retarded development of XO conceptuses during early pregnancy in the mouse. *Journal of Reproduction and Fertility*, **68**, 387–93.

Cole, W.A. & Trasler, D.G. (1980). Gene–teratogen interaction in insulin induced mouse exencephaly. *Teratology*, **22**, 125–39.

Falconer, D.S. (1955). Patterns of response in selection experiments in mice. *Cold Spring Harbour Symposium on Quantitative Biology*, **20**, 178–96.

Falconer, D.S. (1973). Replicated selection for bodyweight in mice. *Genetical Research, Cambridge*, **22**, 291–321.

Falconer, D.S., Gauld, I.K. & Roberts, R.C. (1978). Cell numbers and cell sizes in organs of mice selected for large and small body size. *Genetical Research, Cambridge*, **31**, 287–301.

Falconer, D.S., Gauld, I.K., Roberts, R.C. & Williams, D.A. (1981). The control of body size in mouse chimaeras. *Genetical Research, Cambridge*, **38**, 25–46.

Gauld, I.K. (1980). Prenatal growth and development in fast and slow growing strains of mice. PhD Thesis, University of Edinburgh.

Goss, R.J. (1986). Modes of growth and regeneration: mechanisms, regulation, distribution. In *Human Growth*, ed. F. Falkner & J.M. Tanner, 2nd edition, Vol. 1, pp. 3–26. Plenum Press, New York,.

Gould, J.B. (1986). The low birth-weight infant. In *Human Growth*, ed. F. Falkner & J.M. Tanner, 2nd edition, Vol. 1, pp. 391–413. Plenum Press, New York, London.

Goustin, A.S. Betsholtz, C., Pfeifer-Ohlsson, S., Persson, H., Rydnart, J., Bywater, M., Holmgren, G., Heldon, C.H., Westermark, B. & Ohlsson, R. (1985). Co-expression of the sis and myc proto-oncogenes in developing human placenta suggests autocrine control of trophoblast growth. *Cell*, **41**, 301–12.

Gregg, B.C. (1985). An investigation of the relationship between pattern formation and growth in the mouse vertebral column. PhD Thesis, University of London.

Gregg, B,C. & Snow, M.H.L. (1983). Axial abnormalities following disturbed growth in Mitomycin–C treated mouse embryos. *Journal of Embryology & Experimental Morphology*, **73**, 135–49.

Han, V.KJ.M., D'Ercole, A.J. & Lund, P.K. (1987). Cellular localization of somatomedin (insulin like growth factor) messenger RNA in the human fetus. *Science*, **236**, 193–7.

Healy, M., McLaren, A. & Michie, D. (1960). Foetal growth in the mouse. *Proceedings of the Royal Society, London. Series B*, **153**, 367–79.

Heath, J., Bell, S. & Rees, A.R. (1981). Appearance of funtional Insulin receptors during the differentiation of embryonal carcinoma cells. *Journal of Cell Biology*, **91**, 293–7.

Hill, D.E. (1976). Insulin and fetal growth. In *Diabetes and other endocrine disorders during pregnancy and in the newborn*, ed. M.I. New & R.H. Fiser, pp. 127–39. Alan R. Liss, New York.

Hortsch, M., Schlessinger, J., Gootwine, E. & Webb, C. (1983). Appearance of functional EGF-receptor kinase during rodent embryogenesis. *EMBO Journal*, **2**, 1937–41.

Jones, P.R.M., Peters, J. & Bagnall, K.M. (1986). Anthropometric measures of fetal growth. In *Human Growth*, ed. F. Falkner & J.M. Tanner, 2nd edition, Vol. 1, pp. 251–74. Plenum Press, New York.

Kloosterman, G.J. (1970). On intrauterine growth. *International Journal of Gynecology and Obstetrics*, **8**, 895–912.

Lehtonen, E. & Saxen, L. (1986). Control of differentiation. in *Human Growth*, ed. F. Falkner & J.M. Tanner, 2nd edition, Vol. 1, pp. 27–51. Plenum Press, New York.

Massagué, J. (1987). The TGF–β family of growth and differentiation factors. *Cell*, **49**, 437–8.

McCarthy, J.C. (1965). Genetic and environmental control of foetal and placental growth in the mouse. *Animal Production*, **7**, 347–61.

McKeown, T. & Record, R.G. (1952). Observations on fetal growth in multiple pregnancy. *Journal of Endocrinology*, **8**, 386–401.

McLaren, A. (1965). Genetic and environmental effects on foetal and placental growth in mice. *Journal of Reproduction and Fertility*, **9**, 79–98.

Neligan, G.A., Kolvin, I., Scott, D.McL. & Garside, R.F. (1976). *Born too soon or born too small*. Heinemann Medical Books Ltd, London.

Palmiter, R.D., Brinster, R.L., Hammer, R.E., Trumbauer, M.E., Rosenfeld, M.G., Birnberg, N.C. & Evans, R.M. (1982). Dramatic growth of mice that develop from eggs microinjected with metallo-thionein-growth hormone fusion genes. *Nature*, **300**, 611–5.

Palmiter, R.D., Norstedt, G., Gelinas, R.E., Hammer, R.E. & Brinster, R.L. (1983). Metallothionein-Human GH fusion genes stimulate growth of mice. *Science*, **222**, 809–14.

Popliker, M., Shatz, A., Avivi, A., Ullrich, A., Schlessinger, J. & Webb, C.G. (1987). Onset of endogenous synthesis of epidermal growth factor in neonatal mice. *Developmental Biology*, **119**, 38–44.

Sadler, T.W. (1980). Effects of maternal diabetes on early embryogenesis: I. The teratogenic potential of diabetic serum. *Teratology*, **21**, 339–47.

Seller, M.J. & Perkin–Cole, K.J. (1987). Sex differences in mouse embryonic development at neurulation. *Journal of Reproduction and Fertility*, **79**, 159–61.

Snow, M.H.L. (1983). Restorative growth in mammalian embryos. In *Issues and Reviews in Teratology*, ed. H. Kalter, Vol.1, pp. 251–84. Plenum Publishing Corporation, New York.

Snow, M.H.L. (1986). Control of embryonic growth rate and fetal size in mammals. In *Human Growth*, ed. F. Falkner & J.M. Tanner, 2nd edition, Vol. 1, pp. 67–82. Plenum Press, New York, London.

Snow, M.H.L. (1987). Uncoordinated development of embryonic tissue following cytotoxic damage. In *Approaches to Elucidate Mechanisms in Teratogenesis*, ed. F. Welsch, in press. Hemisphere Publishing Corporation, New York.

Snow, M.H.L. & Gregg, B.C. (1986). The programming of vertebral development. In *Somites in Developing Embryos*, ed. R. Bellairs, D.A. Ede & J.W. Lash, pp. 301–11. Plenum Press, New York, London.

Snow, M.H.L. & Tam, P.P.L. (1979). Is compensatory growth a complicating factor in mouse teratology? *Nature*, **279**, 555–7.

Snow, M.H.L., Tam, P.P.L. & McLaren, A. (1981). On the control and regulation of size and morphogenesis in mammalian embryos. In *Levels of Genetic Control in Development*, ed. S. Subtelny, pp. 201–17. Alan R. Liss Incorporated, New York.

Spiers, P.S. (1982). Does growth retardation predispose the fetus to congenital malformation. *Lancet*, February 6, pp. 312–4.

Tam, P.P.L. (1981). The control of somitogenesis in mouse embryos. *Journal of Embryology and Experimental Morphology*, Supplement 65, 103–28.

Tam, P.P.L. 7 Snow, M.H.L. (1981). Proliferation and migration of primordial germ cells during compensatory growth in mouse mebryos. *Journal of Embryology and Experimental Morphology*, **64**, 133–47.

Tsunoda, Y., Tokunaga, T. & Sugie, T. (1985). Altered sex ratio of live young after transfer of fast- and slow-developing mouse embryos. *Gamete Research*, **12**, 301–4.

Twardzik, D.R. (1985). Differential expression of transforming growth factor α during prenatal development of the mouse. *Cancer Research*, **45**, 5413–6.

Watts, E.S. (1986). Evolution of the human growth curve. In *Human Growth*, ed. F. Falkner & J.M. Tanner, 2nd edition, Vol.1, pp. 153–66. Plenum Press, New York.

VERNON FRENCH

The control of growth and size during development

Introduction

The size and proportions of an animal are reliable morphological features but rather little is understood about the ways in which they are achieved through the control of cell proliferation and cell size throughout development. Typically, each organ and each part of an organ starts to form in its appropriate position at a particular developmental stage, follows a characteristic pattern of growth and, in many animals, stops growing at a predictable time and with a reliable final size.

Spatial organisation is not completely preformed in the egg, although in most cases the cytoplasm is heterogeneous and this may determine the location and orientation of the embryo. The results of experimental embryology show that the detailed patterns of cell fate in most (perhaps all) embryos are formed gradually, through interactions by which cells acquire 'positional information' (Wolpert, 1971), leading them to develop as befits their location. These interactions can be studied by altering the spatial arrangements within the developing embryo, depriving cells of their normal neighbours or placing them against cells which are normally far away. The ensuing alterations in cell fate are often associated with changes in proliferation, and give information on the normal processes whereby cells acquire fates and rates of division appropriate to their positions in the undisturbed embryo.

Pattern formation and growth control have been extensively studied in the *insect epidermis* which, being the surface layer of cells secreting the cuticle, determines the overall size and surface features of the animal. Insect epidermis, particularly that of the legs and wings, is especially suitable for studies of growth control because it is amenable to a wide range of surgical and genetical techniques and there are many mutations altering pattern and growth in specific ways.

I shall concentrate on experimental studies (particularly of the legs of cockroaches and the wings of *Drosophila*) which show a close coordination between pattern regeneration and local growth, and which indicate that the interactions mediating the regeneration of missing parts are normally involved in the patterning and *intrinsic* control of growth within the tissue. I shall then discuss briefly the relevance of these findings for the mechanisms of growth control in vertebrates, where most research has focussed on the actions of circulating hormones and growth factors in regulating cell proliferation, and on the effects of more general factors such as the level of nutrition. There are some indications that these *extrinsic* controls act to modulate rates of growth which are initially controlled much more locally, as in insects, by cellular interactions within developing tissues (Bryant & Simpson, 1984).

Growth and pattern formation in insect appendages

In most insects the thoracic segments form protruding limb buds at an early embryonic stage, and these grow out and become segmented. In the late embryo the epidermis secretes cuticle bearing the characteristic bristles, spines and claws of the larval leg (Figure 1). In each larval stage, or instar, the epidermal cells increase in size, separate from the cuticle, undergo a period of cell division and then secrete a new, larger cuticle. Size increases by a characteristic increment between instars until the adult stage is reached (see Figure 9). In some holometabolous insects such as *Drosophila*, the prospective adult thoracic cells invaginate in the late embryo to form internal imaginal discs (e.g. the mesothorax has a leg disc and a wing disc on each

Figure 1. The larval cockroach leg shown in anterior view and in transverse section through the femur. Five proximal–distal levels (*A–E*) are marked down the tibia. Twelve positions (*1–12*) are marked around the femur circumference, as are the anterior (*antt*), posterior (*post*), medial (*med*) and lateral (*lat*) faces of the leg. The positions of rows of bristles and spines are indicated.

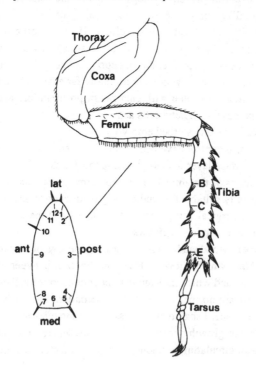

side) which grow but do not form cuticle in the larval instars. The mature discs evaginate, fuse together and secrete the pupal and adult cuticle, forming the appendages and surrounding thorax of the adult.

The larval legs of many insects and the imaginal discs of *Drosophila* will regenerate if damaged, and the interactions governing the process can be studied in a range of grafting and extirpation experiments.

Regeneration in insect leg

If the mid section of a larval leg segment is removed and the distal level is grafted back onto the proximal level stump, local growth occurs at the junction and the intervening mid segment is replaced by *intercalary regeneration* (Figure 2A), restoring the segment to approximately normal size and pattern (Bohn, 1970; Bullière, 1971). If the two levels are confronted by grafting the proximal level onto a distal level stump (Figure 2B), the mid segment is again formed, but now in reversed orientation. The resulting abnormally-long segment has no positional discontinuities and the different regions continue to grow at a normal rate in subsequent instars (Bohn,1970).

If a longitudinal strip of epidermis and cuticle is removed from the leg circumference, cells from different positions heal together and intercalary regeneration restores normal size and pattern (Figure 2C). Similarly, if a strip is moved to a different circumferential position normally non-adjacent cells interact along both graft-host junctions (Figure 2D) and intercalary regeneration forms the intervening section of circumference, by the shortest route (French, 1978, French, Bryant & Bryant, 1976). As in the proximal-distal case, the enlarged circumference has no positional discontinuities and grows on at a normal rate. Histological examination of grafted legs shows that intercalation occurs by localised cell division at the sites of positional discontinuity before and during the period of the instar when the dispersed divisions of normal growth are occurring (Anderson & French, 1985).

After various grafting operations or extensive damage the leg can regenerate supernumerary branches (Bohn 1965, Bullière, 1970, Bart, 1971). For example, if the distal part of one leg is grafted onto the proximal part of the contralateral leg, supernumeraries are formed at the two positions on the graft-host junction where cells from opposite circumferential positions are confronted (see French et al., 1976, Bryant, French & Bryant, 1981 for details). As in the previous cases, the resulting structure has no positional discontinuities (Figure 3) and all parts continue to grow in subsequent instars.

Studies of the regeneration of larval insect legs and *Drosophila* imaginal discs (see below) have suggested the Polar Coordinate Model (see French et al., 1976), which proposes that epidermal cells of an appendage have stable positional properties (positional values) characteristic of their position in the proximal-distal axis and around the circumference (*A–E* and *1–12* in Figure 1). These positional values direct

Figure 2. Intercalary regeneration

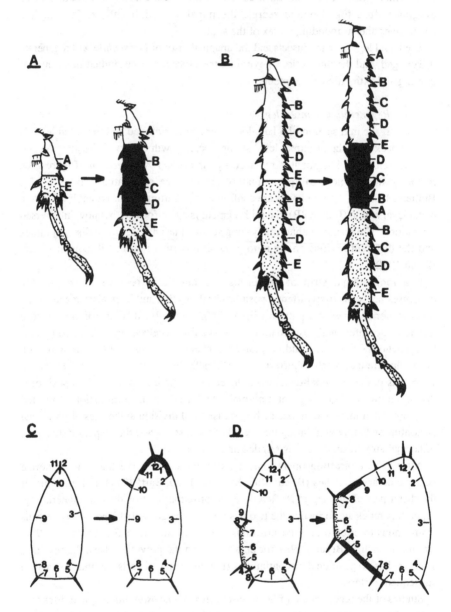

Figure 3. Graft at tibial level between left and right cricket legs, reversing the anterior–posterior (*ant–post*) axis. Circumferential positions *3,6,9* are shown on the graft and host tissue. Shortest route intercalation will occur at the junction, and complete circumferences will form anteriorly and posteriorly, leading to regeneration of supernumeries (see Bryant *et al.*, 1981), for details) and removal of positional discontinuities.

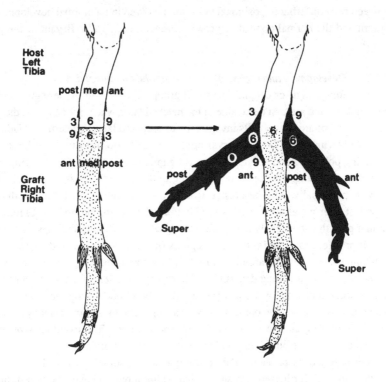

Caption to Fig. 2 (contd).

(A, *B*) interaction between proximal (*A*) and distal (*E*) levels of the cockroach tibia leads to local growth and the formation of an intercalary regenerate consisting of the mid levels (*B,C,D*).

(*C*) After removal of a strip of lateral fumur epidermis, cells from circumferential positions *11* and *2* are confronted and local growth leads to the replacement of cells with positional values *12* and *1*.

(*D*) The graft of medial face of left femur to the anterior face of the host left femur results in intercalary regeneration of the shortest section of circumference separating graft and host positions (i.e. *5, 6, 7*) and not *3,2,1,12,11,10,9* form between positions *4* and *8*).

Graft tissue is stippled and the regenerate is shown in black.

the formation of appropriate cuticular structures and the interaction between cells with different values provokes local growth, producing new cells with intermediate values to form an intercalary regenerate (and, in some cases, a distal regenerate – see Bryant *et al* , 1981 for details).

Work on *Drosophila* imaginal discs suggest that the same sorts of interactions between cells differing in positional value are involved in the normal development of pattern and the normal control of growth (French *et al.*, 1976, Bryant & Simpson, 1984).

Development and regeneration in Drosophila imaginal discs

Imaginal discs originate as small groups of cells within the segment of the early embryo and are first detectable in the hatched larva when, for example, the wing disc consists of around 38 cells invaginated beneath the larval epidermis (Madhavan & Schneiderman, 1977). The larval epidermal cells do not divide in larval life, becoming polyploid and polytene, increasing greatly in size and finally dying in the puparium. The wing disc cells start dividing in the late first instar and the disc then grows exponentially until third instar, when growth rate gradually falls. Cell division stops just after pupariation, with a final cell number of around 50,000 (Figure 4A). Similar growth curves apply to the other imaginal discs (Bryant & Simpson, 1984).

Cell proliferation within different regions of the disc can be studied by several techniques. S phase cells are unevenly distributed and some regions cease division before the rest of the wing disc (O'Brochta & Bryant, 1985). Cell division can also be monitored indirectly by clonal analysis whereby mitotic recombination is induced at a particular time in development, causing a single cell to become homozygous for a mutation altering cuticle colour or bristle morphology. The eventual size of the marked clone on the adult wing indicates the number of divisions undergone in the founder cell and its progeny. This technique shows some regional differences in growth rate and, furthermore, it shows that within a region the exact pattern of cell lineage is not fixed, since clones can occupy partially overlapping positions in different individuals and can vary considerably in size (Figure 4B). Moreover, in mosaics in which only one cell is wild type and the rest are slow-growing *Minute*, the marked cell will produce an extremely large clone but the resulting structure will still be normal in overall size and pattern (see Bryant & Simpson, 1984).

Drosophila only achieves 'normal' size if kept in optimal conditions; in crowded cultures most adults are smaller and very variable in size, and controlled feeding experiments have shown that this is due to the level of nutrition in late larval life. If larvae are removed from the food before they have reached a 'critical weight' (corresponding to early third instar) they survive for a long time but fail to pupariate, while larvae removed later will pupariate on time as small larvae and will produce small adults (Robertson, 1963). In general, variation in nutrition after early third instar will affect adult size but not pupariation time. Although there have been no direct studies on disc cell division in partially starved larvae, the small adults have been reported to have reduced cell number and cell size (Simpson, 1979; Held, 1979). An early triggering of pupariation has also been demonstrated in experiments using temperature sensitive (ts) mutants which inhibit cell division or cause death of disc cells. Pulses of restrictive temperature before early third instar delay development but do not affect size, while later pulses do not delay pupariation but result in small flies (Simpson & Schneiderman, 1976), suggesting that the state of disc development in early third instar (rather than larval weight per se) triggers the hormonal changes eventually leading to pupariation.

The slowing and cessation of cell division in the discs of undamaged, well-fed larvae coincides with pupariation but there is now considerable evidence that it does not depend on the associated hormonal changes. In experiments where early temperature pulses were given to larvae mosaic for wild type and ts mutant tissue, some animals did not delay development and those structures derived from mutant discs were reduced in size. However other animals delayed pupariation, their mutant structures were of normal size but their wild type structures were *not* enlarged, indicating that these discs had not grown beyond their normal cell number and cell

Figure 4. Development of the *Drosophila* imaginal wing disc.

(A) Increase in wing disc cell number from hatching (*H* – 24 h after oviposition) to pupariation (*P* – approximately 120 h) – data redrawn from Bryant & Levinson, 1985). There are three larval instars, separated by moults (*M*). The mature larva is shown with the imaginal discs and histoblast nests (stippled) which will form the adult epidermis.

(B) Adult fly showing the left wing and dorsal mesothorax (black) derived from the left wing disc. Marked clones drawn on the right wing were genrated in different animals by X-ray induced mitotic recombination of 48 h larvae – clones redrawn from Bryant & Simpson, 1984. Partially overlapping clones (*a,b*) show that cell lineage is not defined in disc development. Cells of clone *c* had a higher growth rate than the rest of the disc but the resulting wing is normal in size and structure.

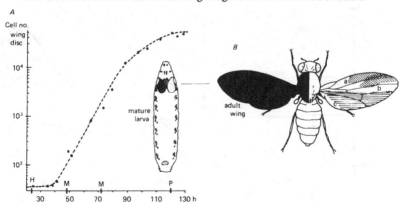

size, despite the extension of larval life (Simpson, Berreur & Berreur-Bonnenfant, 1980).

Autonomous termination of cell division is shown directly by experiments where discs are dissected out of mature or immature larvae and cultured for several days *in vivo* in the abdomens of adult flies. In these conditions undamaged mature discs will not increase in cell number or in total volume (Adler, 1981), but the cell number in immature discs or in a large fragment of a mature disc will slowly increase up to (but *not* beyond) the normal final number (Adler, 1981; Bryant & Levinson, 1985), as shown in Figure 5. These experiments suggest that there is a 'target cell number' (and probably a maximum cell size) which is achieved (given permissive hormonal conditions and adequate nutrition) but is not exceeded. The behaviour of regenerating disc fragments demonstrates, however, that it is not total cell number *per se* but the completion of a normal arrangement and spacing of positional values which leads to growth termination (Bryant & Simpson, 1984).

If a mature wing disc is dissected out, a 1/4 sector removed and the large fragment reimplanted into a mature host larva, it will differentiate at metamorphosis into a partial pattern of wing and thorax structures. By this means a fate map can be constructed, showing which adult structures will form from particular regions of the disc (Bryant, 1978). However, if the 3/4 fragment is given a few days in an adult

Figure 5. Cell number in growing and regenerating wing discs.
(A) Intact discs from mature (5 day) larvae do not increase in cell number during *in vivo* culture in the adult abdomen but, during regeneration, a large fragment re-establishes normal cell number – data redrawn from Adler, 1981.
(B) Increase in cell number in immature (4 day) discs left *in situ* or cultured *in vivo*. Wild type discs slowly attain normal cell numbers *in vivo* . In the epithelial overgrowth mutant (*l(2)gd* (dashed lines) pupariation is delayed or prevented and disc cell number continues to increase. Mutant discs also continue to grow during extended *in vivo* culture. Data redrawn from Bryant & Levinson, 1985.

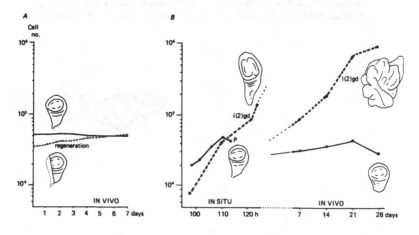

abdomen before being reimplanted into a host larva, it differentiates into a complete pattern (Haynie & Bryant, 1976), while small 1/4 fragments duplicate their fate map structures. By a variety of techniques, regeneration of disc fragments has been shown to be comparable to intercalation in larval legs (see Figure 6), resulting from local cell division at the positional discontinuity at the cut and healing edges (Dale & Bownes, 1980; Keihle & Schubiger, 1985; O'Brochta & Bryant, 1987). It seems that cell division continues until the normal (or the abnormal but duplicated) pattern of positional values is established and then ceases, although extended culture of duplicating fragments does sometimes give continued growth and regeneration of other parts (Kirby, Bryant & Schneiderman, 1982). In contrast, if the disc is thoroughly disrupted by mincing, multiple sites of discontinuity will be created, explaining the great stimulation of growth and regeneration of multiple copies of pattern elements (see Bryant, 1978).

The patterns of cell division in disc development and regeneration support the suggestion (French *et al.*, 1976; Bryant & Simpson, 1984) that the embryonic disc (or other region of insect epithelium) is established initially as a few cells bearing very different positional values. The cell divisions of normal growth are stimulated by the discontinuities between adjacent cells and serve to intercalate intermediate values until the map is sufficiently fine-grained that the small discontinuities no longer provoke cell division (Figure 7). This model explains the sigmoid form of growth seen in discs (as in Figure 4A) and many other organs (Bryant & Simpson, 1984). At early stages there will be discontinuities everywhere so growth depends on the number of cells present (it is exponential), while at later stages the discontinuities are gradually reduced and growth slows down and ceases.

In *Drosophila*, mutations at many loci disrupt growth in a variety of ways, and their genetic and molecular analysis may eventually reveal the role of particular gene products in pattern formation and intrinsic growth control (see Bryant, 1987). Many of these mutations prevent or delay termination of growth in the imaginal discs, giving a greatly increased cell number. Pupariation is delayed or prevented, probably as a result of the effect on the discs, since disc overgrowth has proved to be autonomous when mutant discs are transplanted to wild type larvae. In some of the mutants epithelial organisation breaks down and the tissue no longer resembles an imaginal disc, but in the 'epithelial overgrowth' class (e.g.*l(2)gd* – Bryant & Schubiger, 1971; Bryant and Levinson, 1985) the disc becomes much enlarged, often showing clear duplications in the characteristic patterns of folding (Figure 6B). *l(2)gd* discs implanted into the adult abdomen grow steadily over 28 days to attain some twenty times the final cell number reached in wild type discs (Bryant & Levinson, 1985). Patterns differentiated in epithelial overgrowth mutants, either in animals which eventually pupariate or after disc transplants, frequently show missing or disrupted areas (especially in the leg tarsi) and also areas where pattern elements

Figure 6. Diagrammatic representation of regeneration in the wing disc (based on illustrations in O'Brochta & Bryant, 1987 and Bryant, 1987)

(A) The mature wing disc is dissected out, a 90°-sector removed and cells near the edges marked by intracellular injection of rhodomine dextran (S. Fraser & P. Bryant, unpublished).

(B) After 1 day of *in vivo* culture the wound has closed (although there is usually only temporary healing of epithelium to peripodial membrane) and there is a localised band of cells in S phase (stippling).

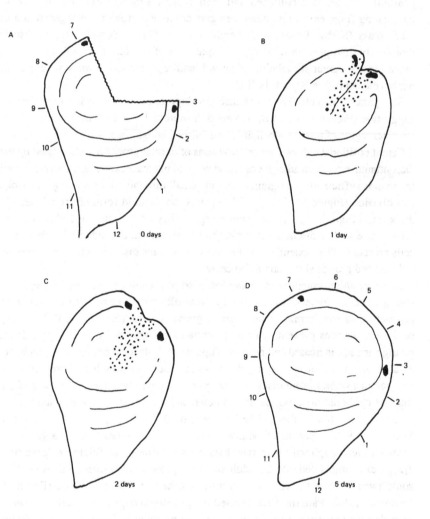

are duplicated (Bryant & Schubiger, 1971; Bryant, 1987). The phenotypes suggest that these mutations disrupt in various ways the co-ordination between pattern formation and cell division, perhaps by causing the cells to lose their ability to interact and hence to control their division and the positional values of their progeny in accordance with those of their neighbours.

> Figure 7. Cell division and change of positional value during intercalary regeneration and normal growth. The rows of boxes represent the epidermis, with the complete patern indicated by positional values *1–10*. Cells stimulated to divide are marked with a dot. There many possible rules for division and change of values; here most cells experiencing a discontinuity divide, giving one daughter with unchanged positional value and the other with a value roughly midway between the original and that of the neighbour. In regeneration at a large discontinuity (arrow), division will be initiated at the junction and will continue within the new tissue until no discontinuities remain. In normal growth of immature tissue with few positional values, divisions are dispersed and numerous at early stages, and gradually become less frequent and stop as discontinuities of more than one unit gradually disappear. If cells are sensitive to smaller discontinuities, division will continue until the structure is larger and the positional map more fine-grained.

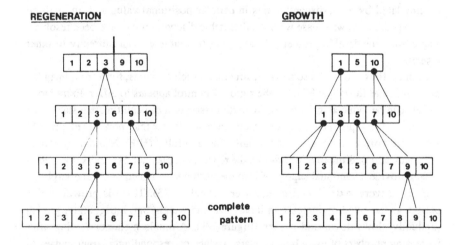

Caption to Fig. 6 (contd).

(*C*) By 2 days healing is complete within the disc epithelium (faint dahsed lines), a band of new cells separates the marked cells and a localised growth zone persists.
(*D*) By 5 days the original marked cells are widely separated and, in 50% of discs, cell divisions have ceased. If the disc is injected into a larva it differentiates to form the complete set of structures, showing that the extirpated portion has been regenerated. In (*A*) and (*D*) circumferential positional values are indicated (equally spaced for simplicity), as in the larval leg in Figure 1. Intercalary regeneration occurs by the shortest route (hence an isolated small *6,5,4,3* fragment will duplicate – see French *et al.*, 1976).

An increase in adult size associated with a normal pattern of differentiation occurs in the mutant, *giant*, and appears to result from an extended larval period caused by a defect in hormone synthesis (Schwartz, Imberski & Kelley, 1984). *Giant* flies appear to have normal cell numbers but an increased *cell size* (Simpson & Morata, 1980). Large adults also result if development occurs at low temperature but, again, this seems to result only from an increase in cell size (Robertson, 1959). An increase in adult size can be achieved through artificial selection and, in lines selected for thorax length (L. Partridge & K. Fowler, unpublished) there is an associated increase in the size of other parts. Preliminary studies on the adult wing blade indicate an increased *cell number* (L. Partridge & V. French, unpublished), so this could be a case where the degree of positional value discontinuity required to stimulate cell division has been altered.

Growth control in other insects

The work on growth and regeneration discussed above suggests that growth control in *Drosophila* is intrinsic to the imaginal discs. Cell division is a component of the mechanism of pattern formation, can occur from early larval stages to pupariation, is stimulated by local discontinuities in cellular positional value and, in well-fed undamaged larvae, will cease when all detectable discontinuities have been resolved. There are considerable problems in applying this simple model directly to other insects.

In butterflies and moths some adult structures, such as wings, form from imaginal discs, much as in *Drosophila*, but their growth control appears to differ. Extra larval moults can be induced in *Manduca* by partial starvation of penultimate instar larvae (Nijhout, 1975), producing larvae which are much larger than normal (Figure 8A) and metamorphose to give abnormally large adults (H.F. Nijhout, personal communication). Similarly, in *Galleria*, the repeated implantation of endocrine organs which provide a continual supply of juvenile hormone can postpone metamorphosis and give several extra larval moults (see Sehnal, 1985). There is growth in this period, as indicated by increase in larval size and weight but, intriguingly, growth gradually slows down and ceases (Figure 8B). Animals metamorphosing after increasing numbers of extra larval instars produce correspondingly larger pupae. At present it is not known that the extra growth involves the continuation of cell division, but the results so far do suggest that epidermal cell divisions in Lepidoptera may be provoked by local cell interactions, as in *Drosophila*, but that they are normally curtailed by metamorphosis (resulting from a fall in juvenile hormone level) long before positional value discontinuities have been resolved.

If normal growth and intercalary regeneration are similar responses to positional value discontinuities (Figure 7), then it is not clear why many insects cannot regenerate. The locust, for example, follows a reliable growth curve to produce adults of predictable size, but is unable to regenerate after leg amputation or to show detectable stimulation of growth following leg grafting operations which associate different proximal-distal or circumferential positions (French, unpublished). Most insect epidermis, unlike that of imaginal disc, can only divide during a particular phase of each moult cycle. It is possible that growth in 'non-regenerating' insects is provoked by local discontinuities but the epidermis is unable to produce any extra cell divisions in this period, even in response to a very large positional discontinuity created between adjacent cells.

In regenerating insects, such as cockroaches and crickets, a large discontinuity clearly does provoke local cell divisions, but this can have effects elsewhere in the animal. Amputation stimulates cell division in the distal region of the stump, producing a small regenerated leg which continues to grow at an elevated rate for one

Figure 8. Induction of supernumary larval moults and extra growth in Lepidoptera.

(A) Development in *Manduca sexta*, showing the growth (measured by head capsule width) between the fourth (*IV*) and the final fifth (*V*) larval instar, and then metamorphosis (double arrows) to pupa (*P*) and adult (*Ad*). The range of larval sizes is shown by heavy bars. After partial starvation (*S*) of the fourth instar, the fifth (*V*) is small and of variable size (dotted bar). Larvae below 5.1 mm in cranial width (*t*) mult into sixth instar larvae (*VI'*), again of very variable size, including very large animals which metamorphose into abanormally large adults (*Ad'*). Redrawn from Nijhout, 1975).

(B) Normal and prolonged development in *Galleria mellonella*, redrawn from Sehnal, 1985. The solid line shows growth in head capsule width up to the seventh instar (*VII*) which normally metamorphoses. Repeated endocrine gland implantations prolong larval development (dashed lines) and growth, producing some very large eleventh instar larvae (*XI'*) and pupae.

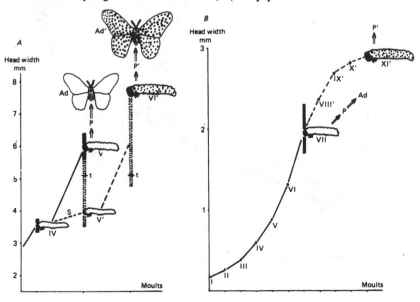

or two further instars before it resumes a normal growth rate (Figure 9). However it remains smaller than a control leg of the same instar, and the animal frequently has a supernumerary larval instar (Bullière, 1968). The formation and growth of the regenerated leg is associated with a dramatic *decrease* in growth of the rest of the epidermis in the operative instar and a decreased growth rate for several subsequent instars, so that the regenerate and contralateral legs become matched in size (Figure 9). There could be some form of direct competition, with the regenerate depriving other epidermis of nutrients, but this seems unlikely in well-fed animals, and the mechanism of this modulation of growth is currently unknown.

Figure 9. Growth of legs during regeneration in the cockroach – redrawn from Bullière, 1968. The graph shows increase in the metathoracic tibia length in normal development (*N*) and in the regenerated (*R*) and contralateral (*C*) legs following an amputation (*amp*) between femur and trochanter early in the first (*a*) or sixth (*b*) larval instar. The schematic drawings show the sequence from operation to first post-operative instar with the small regenerate (*R*) and little growth of the contralateral leg (*C*), to the final result with the two legs of similar size.

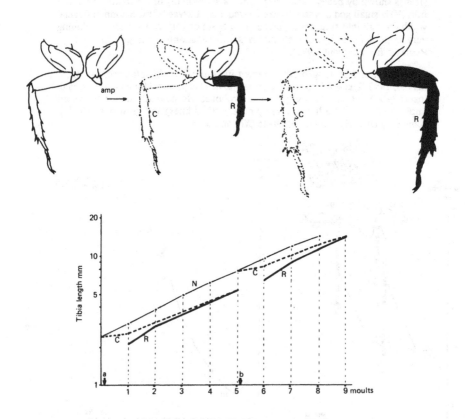

Intrinsic control of growth in vertebrates

Experimental studies on developing insects clearly demonstrate that growth control, at least in the epidermis, is integrated with the development of patterns of cell differentiation. There is considerable evidence for a similar intrinsic control of growth in vertebrate development (reviewed by Bryant & Simpson, 1984) and this is perhaps best illustrated by studies on the limb.

Amphibian limb buds grafted to hosts of different ages or onto host species bearing differently sized limbs generally continue to develop at their normal rate and eventually form a limb characteristic of the donor in size and pattern (Twitty & Schwind, 1931). Parts of chick and mammalian limbs have been shown to grow on at normal rates and achieve normal sizes when transplanted elsewhere in the limb, onto the body or, in some cases, when explanted into culture (see Bryant & Simpson, 1984). These results show that there must be a basic intrinsic control of limb growth, although in some circumstances this can be modulated dramatically by extrinsic factors (see below).

Experimental studies on chick and amphibian limb development suggest that, as in insects, intrinsic growth control is integrated into the mechanism of pattern formation. In the early chick embryo, the successive formation of the more distal limb parts depends on cell proliferation in the apical region of the limb bud. Furthermore, grafting experiments which confront anterior and posterior cells lead to a duplication of the pattern and a stimulation of cell division, initially around the discontinuity and subsequently over a wider area (Cooke & Summerbell, 1980). The results do not suggest an interaction as local as that seen in insect epidermis, but do argue that related mechanisms control the pattern of differentiation and rate of cell division in the early limb (Summerbell, 1981).

In a variety of grafting experiments on the buds and larval legs of amphibians normally non-adjacent cells are confronted, and intercalary or supernumerary regeneration occurs to remove discontinuities in the pattern, much as in the insect epidermis (French *et al.*, 1976; Bryant *et al.*, 1981). It has not been shown directly that regeneration in amphibian limbs results from local stimulation of cell division, but the close similarity to the insect results suggests that the relationship between positional values of neighbouring cells may well control cell division during the formation of the vertebrate limb, in its response to surgery or damage, and perhaps in the growth of the intact limb to maturity.

The ability of the limb and other vertebrate organs to execute the 'normal' pattern of growth depends upon adequate nutrition, as in insects, and at least the mammals have the ability to increase growth rate to return to a normal growth profile after periods of starvation (see Tanner 1981; Mosier, this volume). Growth rate can also be modulated directly by hormone levels, as in the dramatic increase and decrease seen in post-embryonic mice and humans with elevated or depressed levels of growth hormone (see Bryant & Simpson, 1984).

There has been much recent research on the effects on cell proliferation (usually *in vitro*) of a variety of growth factors, usually isolated from tissues or cultured cell lines. The role of these factors in growth control *in vivo* is not yet well understood, and it is possible that at least some of them could normally act locally within tissues, being retained on the cell surface or stable over only a short distance, rather than being endocrine agents (see Clemmons, Hill, this volume). In this case molecules currently recognised as 'growth factors' could be part of intrinsic mechanisms of growth and pattern control (Smith, 1981). It is perhaps relevant that there are now several cases of great structural similarity between a vertebrate *growth* factor and the putative product of a *Drosophila* gene whose mutant alleles disrupt *pattern* (see Bryant, 1987).

Growth within the tissues of vertebrates (and other animals) will surely prove to be controlled by a complex interplay between intrinsic interactions (linking the size of an organ to the pattern of cell fates developing within it) and extrinsic signals (modulating the growth of the animal, or of a specific organ, in accordance with its nutritional and hormonal state).

Acknowledgement
It is a pleasure to thank Peter Bryant and Linda Partridge who have informed, suggested and criticised.

References
Anderson, H. & French, V. (1985). Cell division during intercalary regeneration in the cockroach leg. *Journal of Embryology and Experimental Morphology*, **90**, 57–78.

Adler, P. (1981). Growth during regeneration in imaginal discs. *Developmental Biology*, **87**, 356–73.

Bart, A. (1971). Morphogenèse surnuméraire au niveau de le patte du Phasme *Carausius morosus*. *Wilhelm Roux's Archives*, **166**, 331–64.

Bohn, H. (1965). Analyse der Regenerationsfähigkeit der Insekten-extremität durch Amputations- und Transplantationsversuche an Larven der afrikanschen Schabe *Leucophaea maderae* Fabr. (Blattaria). II. Mitt. Achsendetermination. *Wilhelm Roux's Archives*, **156**, 449–503.

Bohn, H. (1970). Interkalare Regeneration und segmentale Gradienten bei den Extremitäten von *Leucophaea*-Larven (Blattaria). I. Femur und Tibia. *Wilhelm Roux's Archives*, **165**, 303–40

Bryant, P.J. (1978). Pattern formation in imaginal discs. In *The Genetics and Biology of* Drosophilia, ed. M. Ashburner & T.R.F. Wright. Volume 2c, pp. 229–335. Academic Press: New York.

Bryant, P.J. (1987). Experimental and genetic analysis of growth control in *Drosophila* imaginal discs. In *Genetic Regulation of Development*, ed. W.F. Loomis, pp. 339–?, 45th Symposium of The Society for Developmental Biology.

Bryant, P.J. & Levinson, P. (1985). Intrinsic growth control in the imaginal primordia of *Drosophila*, and the autonomous action of a lethal mutation causing overgrowth. *Developmental Biology*, **107**, 335–63.

Bryant, P.J. & Simpson, P. (1984). Intrinsic and extrinsic control of growth in developing organs. *The Quarterly Review of Biology*, **59**, 387–415.

Bryant, P.J. & Schubiger, G. (1971). Giant and duplicated imaginal discs in a new lethal mutation of *Drosophila melanogaster*. *Developmental Biology*, **24**, 233–63.

Bryant, S.V., French, V. & Bryant, P.J. (1981). Distal regeneration and symmetry. *Science*, **212**, 993–1002.

Bullière, D. (1968). Étude de le régénération chez un insecte Blattoptéroïde *Blabera craniifer* Burm. (Dictyoptère). II. Influence du moment de l'amputation dans l'intermue sur la régénération de la patte métathoracique. *Bulletin de la Societé Zoologique de France*, **93**, 69–82.

Bullière, D. (1970). Interprétation des régénérats multiples chez les Insectes. *Journal of Embryology and Experimental Morphology*, **23**, 337–57.

Bullière, D. (1971). Utilisation de la régénération intercalaire pour l'étude de la détermination cellulaire au cours de la morphogenèse chez *Blabera craniifer* (Insecte Dictyoptère). *Developmental Biology*, **25**, 672–709.

Cooke, J. & Summerbell, D. (1980). Cell cycle and experimental pattern duplication in the chick wing during embryonic development. *Nature*, **287**, 697–701.

Dale, L. & Bownes, M. (1980). Is regeneration in *Drosophila* the result of epimorphic regulation? *Wilhelm Roux's Archives*, **189**, 91–6.

French, V. (1978). Intercalary regeneration around the circumference of the cockroach leg. *Journal of Embryology and Experimental Morphology*, **47**, 53–84.

French, V., Bryant, P.J. & Bryant, S.V. (1976). Pattern regulation in epimorphic fields. *Science*, **193**, 969–81.

Haynie, J.L. & Bryant, P.J. (1976). Intercalary regeneration in the imaginal wing disc of *Drosophila*. *Nature*, **259**, 659–62.

Held, L.I. (1979). Pattern as a function of cell number and cell size on the second leg basitarsus of *Drosophila*. *Wilhelm Roux's Archives*, **187**, 105–27.

Keihle, C.P. & Schubiger, G. (1985). Cell proliferation changes during pattern regulation in imaginal leg discs of *Drosophila melanogaster*. *Developmental Biology*, **109**, 336–46.

Kirby, B.S., Bryant, P.J. & Schneiderman, H.A. (1982). Regeneration following duplication in imaginal wing disc fragments of *Drosophila melanogaster*. *Developmental Biology*, **90**, 259–71.

Madhavan, M.M. & Schneiderman, H.A. (1977). Histological analysis of the dynamics of growth of imaginal discs and histoblast nests during the larval development of *Drosophila melanogaster*. *Wilhelm Roux's Archives*, **183**, 269–305.

Nijhout, H.F. (1975). A threshold size for metamorphosis in the tobacco hornworm, *Manduca sexta*. *Biological Bulletin*, **149**, 214–25.

O'Brochta, D.A. & Bryant, P.J. (1985). A zone of non-proliferating cells at a lineage restriction boundary in *Drosophila*. *Nature*, **313**, 138–41.

O'Brochta, D.A. & Bryant, P.J. (1987). Distribution of S-phase cells during the regeneration of *Drosophila* imaginal wing discs. *Developmental Biology*, **119**, 137–42.

Robertson, F.W. (1959). Studies in quantitative inheritance. XII. Cell size and number in relation to genetic and environmental variation of body size in *Drosophila*. *Genetics*, **44**, 869–96.

Robertson, F.W. (1963). The ecological genetics of growth in *Drosophila*. 6. The genetic correlation between the duration of the larval period and body size in relation to larval diet. *Genetical Research Cambridge*, **4**, 74–92.

Schwartz, M.B., Imberski, R.B. & Kelley, T.J. (1984). Analysis of metamorphosis in *Drosophila melanogaster*, characterisation of *giant*, an ecdysteriod deficient mutant. *Developmental Biology*, **103**, 85–95.

Sehnal, F. (1985). Growth and life cycles. In *Comprehensive Insect Physiology, Biochemistry and Pharmacology*, ed. G.A. Kerkut & L.I. Gilbert, Volume 2, 1–86. Pergamon, Oxford.

Simpson, P. (1979). Parameters of cell competition in the compartments of the wing disc of *Drosophila*. *Developmental Biology*, **69**, 182–93.

Simpson, P., Berreur, P. & Berreur-Bonnenfant, J. (1980). The initiation of pupariation in *Drosophila*: dependence on growth of the imaginal discs. *Journal of Embryology and Experimental Morphology*, **57**, 155–65.

Simpson, P. & Morata, G. (1980). The control of growth in the imaginal discs of *Drosophila*. In *Developmental and Neurobiology of Drosophila*, ed. O. Siddiqi, P. Babu, L.J. Hall & J.C. Hall, pp. 129–39. Plenum, New York.

Simpson, P. & Schneiderman, H.A. (1976). A temperature-sensitive mutation that reduces mitotic rate in *Drosophila melanogaster*. *Wilhelm Roux's Archives*, **179**, 215–36.

Smith, J.C. (1981). Growth factors and pattern formation. *Journal of Embryology and Experimental Morphology*, **65** (Supplement), pp. 187–207.

Summerbell, D. (1981). Evidence for regulation of growth, size and pattern in the developing chick limb bud. *Journal of Embryology and Experimental Morphology*, **65** (Supplement), pp. 29–150.

Tanner, J.M. (1981). Catch-up growth in man. *British Medical Bulletin*, **37**, 233–38.

Twitty, V.C. & Schwind, J.L. (1931). The growth of eyes and limbs transplanted heteroplastically between two species of Amblystoma. *Journal of Experimental Zoology*, **59**, 61–86.

Wolpert, L. (1971). Positional information and pattern formation. *Current Topics in Developmental Biology*, **6**, 183–224.

H.D. MOSIER

Catch-up growth and target size in experimental animals

Introduction

The phenomenon of catch-up growth has long been recognized in the clinic (Prader, Tanner & von Harnack, 1963; Tanner, 1963) and in the laboratory (Hatai, 1907; Osborne & Mendel, 1914) (Figure 1), but the mechanism controlling it remains unknown (see reviews: Tanner, 1979; Tanner, 1981; Van den Brande, 1986). One of the principal unresolved problems in catch-up growth has been the question whether the phenomenon is controlled by a central mechanism or whether it occurs by multicentric mechanisms in peripheral tissues. Tanner (1963) has suggested a model for the central control of catch-up growth which requires a sensor of the degree of growth deficit and a regulator of growth rate according to the deficit. Alternatively,

Figure 1. Results of an early experimental study of catch-up growth. Young rats were fed a diet of cornstarch and water for 21 days beginning at 30 days of age and then fed a normal diet to sacrifice at 200 days of age. Males, continuous lines; females, broken lines. (From Hatai, 1907).

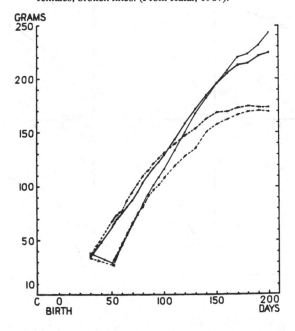

cells may be 'programmed for a certain amount of growth' and simply follow the program if inhibiting factors are removed (Prader, 1978). However, the hypothesis for a central control is attractive for its ability to account for fine-tuning of growth recovery under varying conditions (Tanner, 1981) as well as for offering opportunities for experimental attack.

In the present report we describe experiments which test the hypothesis of central control. In addition, we summarize work which delineates factors limiting catch-up growth.

Experimental models

In most of our experiments we have used the black-hooded Long–Evans strain of rat because it has a larger body size and a relatively vigorous constitution in comparison with other strains of domesticated rats. Fasting has been carried out by withholding food for 24 to 72 hours at 40 days of age with the animals housed in hanging cages to prevent coprophagia. Complete catch-up growth occurs with refeeding. High dose glucocorticoid treatment has been carried out by giving daily subcutaneous injections of cortisone acetate in a dose of 5 mg/rat (approximately 40

Figure 2. Body weight and tail length growth curves of male rats submitted to a period of fasting, cortisone treatment or propylthiouracil (PTU)-induced hypothyroidism. The fasted group caught up completely. The cortisone group failed to catch-up. The PTU group underwent partial catch-up growth. (From Mosier, 1971).

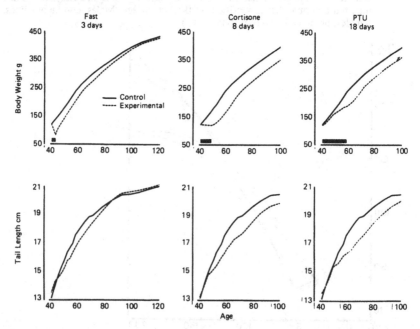

mg/kg) for 4 to 8 days from 40 days of age. This dose suppresses growth without causing significant weight loss. Growth rate of body weight and tail length resumes promptly after cessation of treatment; however, catch-up does not occur through 100 days of age. Hypothyroidism has been induced by feeding 0.1 percent propylthiouracil in stock diet for 13–18 days beginning at about 40 days of age. Only partial catch-up growth occurs during refeeding with normal diet through 100 days of age (Figure 2) (Mosier, 1971).

X-irradiation of only the head of the neonatal rat using a single dose of 3.5 Gy (350 rad) or greater results in permanent stunting of body weight (Yamazaki, Bennett, McFall & Clemente, 1960; Yamazaki, Bennett & Clemente, 1962; Mosier & Jansons, 1967; Sakovic, 1969) and tail length (Mosier & Jansons, 1967). In our experiments the rats have been given a single dose of X-radiation at 2 days of age to the whole head ot to portions of the head (Figure 3). Littermate controls are handled identically but are completely shielded from the beam. After 6.0 Gy (600 rad) head irradiation at 2 days of age body weight velocity is decreased from the second week of life through 107 days of age. Tail length velocity is decreased from the second week of life through 37 days of age (Figure 4). The neuropathology of neonatal head-irradiation in the rat has been well described (Clemente, Yamazaki, Bennett & McFall, 1960; Schjeide, Yamazaki, Haack, Ciminelli & Clemente, 1966). Marked disturbance of neuronal organization and reduction of axons is evident, but the degree of cellular disorganization and cellular deficit differs markedly in different areas of the brain. Differentiation and growth of the cerebellum are more markedly impaired by irradiation at this age than differentiation and growth of the cerebrum (Figures 5a,b, 6).

The set point for body size

The hypothesis of central control of catch-up growth requires a reference or set point for normal body size for age (Tanner, 1963). The stunted head-irradiated rat model was used to test the set point hypothesis.

The effect of head-irradiation on growth is dose related. In the rat, head-irradiation with doses of 2.5 Gy (250 rad) or below do not disturb overall growth, but at dose levels of 3.5 Gy (350 rad) or greater there is dose related stunting as the dose increases to lethal levels 7.5 Gy (750 rad) or greater. Experiments with partial head-irradiation have shown that the growth impairment is not a non-specific effect of tissue damage, for growth stunting follows irradiation only if the irradiation is bilateral and involves midline structures (Figure 7) (Mosier & Jansons, 1970). Food intake relative to body weight is normal both during suckling and after weaning in the stunted head-irradiated rat (Mosier & Jansons, 1967).

Further studies in head-irradiated rats have failed to show disturbances of known endocrine functions. Pituitary concentrations of bioassayable thyrotropin, gonadotropin, and growth hormone (GH) relative to wet weight are not changed significantly (Mosier & Jansons, 1968). Shielding the pituitary from the X-ray beam does not reduce the degree of stunting (Mosier & Jansons, 1970). Treatment of head-irradiated stunted rats with bovine GH and/or thyroxine from ages 30 through 40 days is ineffective in attempts to restore normal growth (Mosier & Jansons, 1967). At 20–21 days of age cell size is reduced significantly in brain, heart and skeletal muscle, and reduced non-significantly in liver and kidney. Cell number is increased in heart and skeletal muscle, and increased non-significantly in brain, liver and kidney (Mosier *et al.*, 1985b). One would expect reduced cell number if hypopituitarism or undernutrition had caused the stunting (Winick & Grant, 1968). Tibial epiphyseal width is normal or increased in stunted head-irradiated rats studies at 70 days of age (Mosier *et al.*, 1983a) instead of narrowed as would be expected in GH deficiency (Evans *et al.*, 1943). Serum bioassayable somatomedin activity is normal in the head-irradiated rat (Wright & Mosier, unpublished findings).

The lack of correlation of stunting with known endocrine functions in the head-irradiated rat contrasts with clinical observations that some children with growth failure after cranio-spinal irradiation for treatment of brain neoplasms respond to GH

Figure 3. Partial head areas that have been irradiated are indicated in solid black shading. The abbreviations conform to those of Figure 7. (From Mosier & Jansons, 1970)

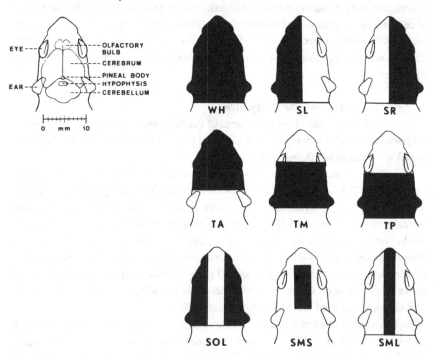

Figure 4. Growth velocity of body weight and tail length in male Long–Evans rats given 6.0 Gy (600 rad) whole-head X-irradiation at 2 days of age and in controls. The velocities are computed on the basis of weekly measurements from 2 days through 121 days of age. Significant differences are as follows: * = $p < 0.05$, ** = $p < 0.005$.

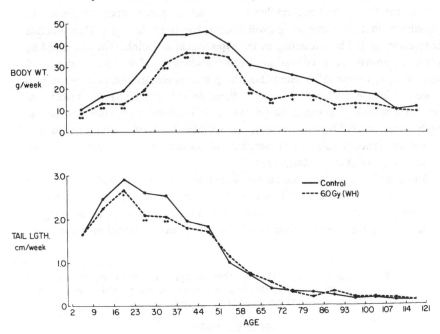

Figure 5. Rat cerebellum at 121 days of age. a. Control. b. Whole head irradiation at 2 days of age with 6.0 Gy (600 rad). Cresyl violet and luxol fast blue.

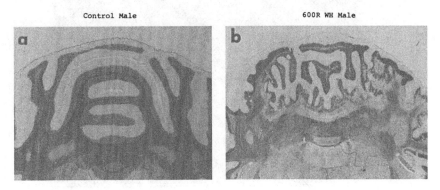

therapy (Shalet, 1986). There are, however, irradiated children who fail to respond to GH therapy (Winter & Green, 1984). These differences may relate to differences between the species in the degree of differentiation of the brain at the time of irradiation, the dose factors, and susceptibility to irradiation injury.

The inability of the head-irradiated rat to recover normal body size raises the possibility that the catch-up growth control has been damaged. One possible mechanism would be a resetting of the hypothetical set-point. That was tested by observing growth in stunted head-irradiated rats which were fasted for 48 hours at 40 days of age and then refed. Normal catch-up growth acceleration occurred in both male and female irradiated and non-irradiated animals (Figure 8). The fasted non-irradiated animals caught up to the non-fasted non-irradiated control rats' size for both body weight and tail length. Similarly, the fasted irradiated rats caught up to the size of the non-fasted irradiated rats' size, but did not catch up to the size of the non-irradiated rats (Figure 9) (Mosier et al., 1983a).

These results indirectly support not only the existence of the set point but its central location. Alternatively, the set point could be located peripherally and be reset as an abscopal effect of head-irradiation. However, that hypothesis has the disadvantage of compounding the existing mystery of abscopal growth retardation in the head-

Figure 6. Rat cerebrum at 121 days after left sagittal half-head (SL) irradiation at 2 days of age with 6.0 Gy (600 rad). A loss of neurons is evident in the gyrus dentatus (arrow). Cresyl violet and luxol fast blue.

600R SL Male

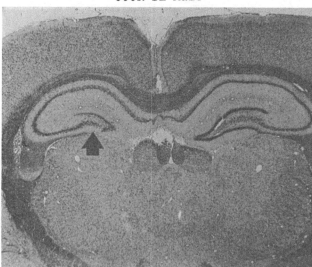

irradiated rat. Uncertainty is likely to remain in this area until central location of the set point can be confirmed by direct methods.

Growth hormone (GH) secretion and catch-up growth

Early studies based on single plasma determinations of GH in rats showed that GH tended to be elevated during recovery after a fast (Sinha *et al.*, 1973; Mosier & Jansons, 1976), after high-dose glucocorticoid treatment (Mosier & Jansons, 1976), and after propylthiouracil-induced hypothyroidism (Mosier *et al.*, 1977).

Metabolic clearance rate (MCR) of GH was determined in the three models in order to exclude a change in MCR as a source of the GH elevations. The results showed that MCR was normal in fasted and PTU-fed rats when plasma GH concentration was elevated during growth recovery. MCR was reduced throughout a 28 day period of observations in the cortisone-treated rats. However, the cortisone rat also had a

Figure 7. Body weight and tail length growth curves of rats receiving 6.0 Gy (600 rad) to the whole head (WH), a sagittal half-head area (SL) or a midline zone (SMS). The zones conform to the patterns in Figure 3. Irradiation limited to the midline zone produces growth effects comparable to those of whole head irradiation. Unilateral irradiation of the brain has no effect on growth. (From Mosier & Jansons, 1970).

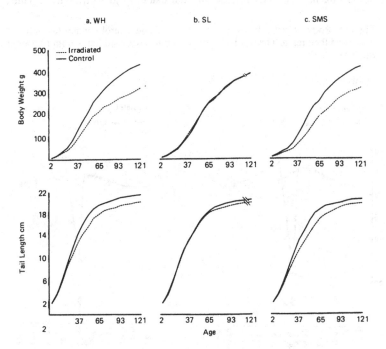

36   H.D. MOSIER

normal MCR to body weight ratio and normal disappearance rates of total and immunoprecipitable label. The reduction of MCR in the cortisone-treated rat was attributed to the reduction of the soft tissue mass and a consequent reduction in number of growth hormone receptors. These findings left open the possibility that increased secretion of GH probably occurred during the recovery phase after all three experimental treatments (Mosier *et al.*, 1980).

Secretory profiles of GH were determined by standard techniques for serial sampling of plasma in cannulated undisturbed rats. Fasted rats, 12 to 26 days into recovery, had, on inspection, apparently normal periodic bursts of GH pulses, with peaks above 200 ng/ml and intervening troughs with concentrations less than 1 ng/ml. The mean period between pulses was lengthened slightly in fasted-refed rats, but the difference was not significant by t test. The area under the curve of GH concentration vs time was significantly greater in fasted-refed rats than in controls. It was also greater in the fasted-refed rats during light, the inactive period of the rat, than during dark; controls did not show this difference (Mosier, Jansons & Dearden, 1985a). A significantly greater area under the GH concentration curve was also found in cortisone-treated rats from 17 to 27 days into recovery (Mosier & Jansons, 1985). These findings linking GH release to growth recovery are compatible with the existence of a link between GH release and the catch-up growth control. Thus,

Figure 8. Body weight velocity in head-irradiated and control females fasted for 48 hours and then refed. The results of the two groups were similar. (From Mosier *et al.*, 1983a).

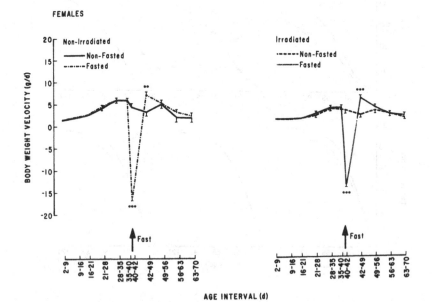

FEMALES

increased secretion of GH in these models may serve as a marker for the positive (switched-on) phase of the catch-up growth control.

GH secretion was studied in the stunted head-irradiated rat in order to validate the hypothesis that irradiation resets the catch-up growth control. Stunted head-irradiated rats, at 47 to 64 days of age, were serially sampled for 18 hour periods, 9 hours in light and 9 hours in dark. In irradiated rats the area under the GH curve was decreased in both the light and the dark periods. Rhythmic secretion of GH appeared to be normal on inspection of the plotted results (Mosier *et al.*, 1985b). The absence of an increase in GH secretion in the stunted head-irradiated rat supports the reset hypothesis. Resetting the set point for a smaller body size would eliminate the necessity for catch-up growth and avoid the sequence that turns on increased GH during catch-up growth.

The mean interval between GH pulses in the fasted-refed rat appeared to be lengthened but the difference was not significant by t test. We have employed time-

Figure 9. Growth curves of body weight of the rats represented in Figure 8 showing complete catch up of both the irradiated and the non-irradiated fasted-refed groups. The irradiated rats caught up only to the size of their normally fed irradiated counterparts. (From Mosier *et al.*, 1983a).

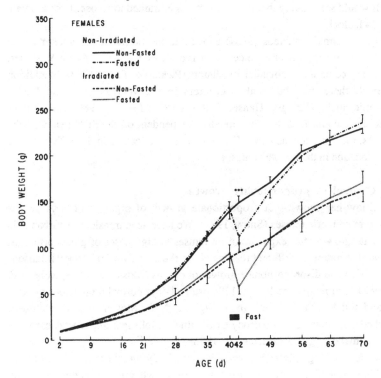

series analysis in order to determine the significance of the change in the frequency. Frequency domain analysis indicated a slower frequency in fasted-refed animals, but the validity of this approach was open to question. Frequency domain analysis is suitable when data are deterministic and sinusoidal, but the GH data have neither characteristic. Although a deterministic rhythm may well exist buried in the data, GH pulses are clearly influenced by a variety of extraneous variables, some of them environmental. In the analysis of GH pulses, therefore, we have used time domain analysis (Box & Jenkins, 1976) with univariate and multivariate techniques. The findings showed, with a high degree of significance, that a cycle of approximately 195 min in length was present in the growth hormone pulses in the fasted-refed rats and was not present in the controls (Figure 10) (H.D. Mosier et al., unpublished).

The functional significance of the 195 min cycle in the fasted group is unclear. It is tempting to speculate that this pulse is part of the mechanism producing the increase of GH secretion that occurs in the recovering fasted rat. The increae in GH secretion during catch-up growth was found only during the lights-on phase of the 24 h sampling period, suggesting entrainment of the catch-up growth mechanism to the 24 h clock (Mosier et al ., 1985a). However, the altered pattern of pulsatile secretion demonstrated by the fasted group appeared to extend through both light and dark periods. It would seem likely that the new cycle is entrained to an oscillator separate from the 24 h clock.

The suprachiasmatic nucleus (SCN) appears to be the major pacemaker in the control of circadian rhythms by the central nervous sytem (Gross, Mason & Meijer, 1983) and may contain two circadian oscillators (Pickard & Turek, 1983). In addition to the SCN, there may be weak oscillators in the ventromedial and lateral hypothalamic nuclei (Krieger, Hanser & Krey, 1977; Inouye, 1982) and another major oscillator coupled to, but functionally independent of, the SCN (Moore-Ede, 1983). Thus, the timer for the new GH rhythm could be located in the SCN itself, or in another location in the hypothalamus.

Control of proportionate growth

Snow has shown that proportionate growth of organs and body size is maintained in chimeric fetuses (Snow, 1981). We have seen transient disturbance in proportionate growth after exposure of rat fetuses to high doses of glucocorticoids. Discordant maturation of different regions of the skeleton (Mosier, Roberts, Jansons & Biggs, 1981) and disproportionate growth of organs (Mosier, Jansons, Roberts & Biggs, 1982) occurred prenatally. By 120 postnatal days normal proportions of body weight and tail length were regained (Fig. 11; H.D. Mosier & L.C. Dearden, unpublished). At that age chondrocytes of tibial epiphyseal and costal cartilage showed minimal changes (Dearden et al., 1986).

Allometric (log–log) plots (Huxley, 1932) of body weight and tail length are disturbed in rats of 40 days of age during transient growth retardation due to fasting, cortisone treatment, or propylthiouracil-induced hypothyroidism; however, during recovery, a prompt return to normal allometry occurs in all three models, even in partial or complete failure of catch-up growth (Figure 12) (Mosier, 1969).

After whole head irradiation of two-day-old rats allometry of body weight and tail length begins to deviate from normal between nine and thirty days of age. Disproportions in body weight/tail length allometry occurred before and after weaning (Figure 13). Eventually, by about 121 days of age normal proportions were attained (Mosier & Jansons, 1971).

Some other conditions may also delay or impede restoration of normal body proportions. Undernutrition during the suckling period in the rat limits the ability of the animal to later achieve full recovery of body proportions (Williams, Tanner & Hughes, 1974). Williams and Hughes showed that proportionate growth of the skeleton and soft tissues of the rat after undernutrition is influenced by the age of rehabilitation (1975) and by sex (1978).

Our observations suggest that there is great pressure within the organism to restore normal body proportions after a growth insult and that restoration and maintenance of

Figure 10. Plasma GH concentration plotted against time in a fasted-refed rat sampled at 15 minute intervals for 24 hours. The darts are spaced at intervals (195 min) corresponding to a frequency found only in the fasted group by a time domain technique. See text.

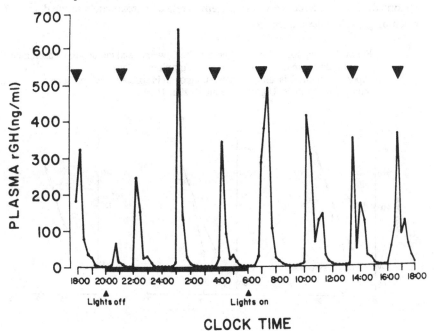

normal body proportions may have priority over controls of unidimensional growth. This seems to indicate that proportionate growth is under an active control. The results from the head-irradiation experiments suggest that the control may be located in the brain.

Factors limiting catch-up growth

Our studies in collaboration with Professor Lyle Dearden have shown that fasting, glucocorticoid treatment, or propylthiouracil feeding are accompanied by alterations in costal or tibial cartilage histology, ultrastructure and histochemistry. Changes extend into the recovery period depending on the treatment. In the fasted-refed rat the cartilage appears nearly normal by 7 days of recovery, although some changes may persist as long as 21 days (Dearden & Espinoza, 1974; Dearden & Mosier, 1974). Cartilage abnormalities persist for longer periods after cortisone treatment. These consist of increased numbers of dead and dying cells, reduction of chondrocyte size, disruption or reduction in the amount of granular and endoplasmic reticulum, and reduction of electron density of the ground substance (Dearden & Mosier, 1972). After propylthiouracil treatment almost complete recovery of chondrocyte ultrastructure occurs by 12 days on a normal diet, but an increase in hydroxyapatite crystals occurs in all zones of the matrix beyond that time (Dearden, 1974; Dearden & Mosier, 1974). The degree of and persistence of ultrastructural abnormalities into the recovery period seem to relate approximately to the degree of catch-up growth in these three models.

Figure 11. Log–log (allometric) plots of body weight and tail length measurements of animals represented in Figure 2. Complete recovery is seen in all three experimental models although catch-up growth is absent in the cortisone group and only partial in the PTU group. (From Mosier, 1969).

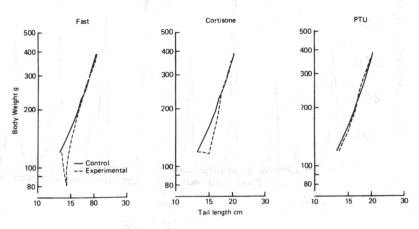

Cartilage metabolism in *in vitro* organ cultures also reflect delays in growth recovery. Incorporation of sulfate by rib cartilage in organ culture is reduced during starvation (Salmon, Bower & Thompson, 1963; Herbai, 1971), glucocorticoid excess (Tessler & Salmon, 1975), and hypothyroidism (Mosier *et al.*, 1977). During recovery after a 2 day fast sulfate incorporation is normal by 7 days and remains normal thereafter (Mosier, Dearden, Jansons & Hill, 1978). After cortisone treatment it returns to normal by 7 days but then exceeds the normal level through at least 28 days of recovery, the limit of the observations (Mosier, Jansons, Hill & Dearden, 1976). After PTU-induced hypothyroidism the pattern is similar to that observed after cortisone (Mosier *et al.*, 1977).

In vitro functions of cartilage of the head-irradiated rat at different ages have been studied just before weaning (20–21 days), just after weaning (23 days), during the rapid postnatal growth phase (40–45 days) and after sexual maturation (70–71 days). Incorporation of labelled sulfate, thymidine, leucine or proline was measured as an

Figure 12. Log–log (allometric) plots of body weight and tail length in rats irradiated with 10.0 Gy (1000 rad, abbreviated R in the figure). The irradiated areas conform to those in Figure 3. The greatest disturbance of proportionate growth occurs in the whole head (WH) group, but normal proportions are eventually regained. The disturbance in the sagittal half-head (SL) group occurred after weaning because of asymmetrical dental development. This was corrected by trimming the incisors and feeding a powdered diet. (From Mosier & Jansons, 1971).

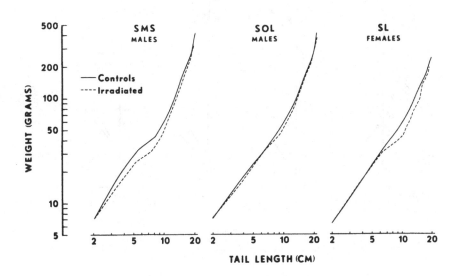

ALLOMETRY AFTER 1000 R

index of the formation of sulfated proteoglycans, DNA, chondroprotein or collagen, respectively. The metabolism of these various precursors showed no significant disturbance at 20–21 days or at 70–71 days. Reduced sulfate and thymidine occurred with weaning and can probably be attributed to the brief period of undernutrition at that time. Increased incorporation of label occurred at 41–45 days of age in cartilage of irradiated rats incubated with labelled sulfate, leucine and proline. Thymidine, on the other hand, did not show an increase at that time, and was even decreased in males (Fig 13; H.D. Mosier *et al.*, 1983b).

The increase in incorporation of sulfate, leucine and proline in the head-irradiated rat occurred when its growth velocity was significantly less than that of controls. The fact that thymidine incorporation was either decreased or not changed indicates that some factor other than somatomedin stimulates proteoglycan, protein and collagen synthesis while restraining DNA synthesis. The differential effects are not characteristic of hypopituitarism. Other possibilities are purely conjectural at this point.

Figure 13. *In vitro* incorporation of labelled substrate in the costal cartilage of stunted head-irradiated rats at 41–45 days of age. The incubations were carried out with and without presence of normal rat serum in the media. See text.

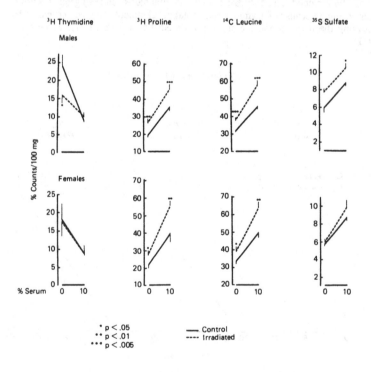

Summary and conclusions

Experimental models of stunted growth in the rat have been employed in studies intended to elucidate the nature of the control of catch-up growth. The models include stunting by neonatal head-irradiation and transient growth retardation produced by periods of fasting, glucocorticoid treatment or hypothyroidism.

The findings in the head-irradiated rat provide evidence that a set point for body size appropriate for age is located in the brain and that the set point can be reset experimentally. Results in the fasted rat indicate that the catch-up growth control is linked to GH release, that the link is entrained to the environmental light-dark cycle, and that another oscillator may influence GH pulses during catch-up growth.

Changes in proportions of body weight and tail length in the head-irradiated rat suggest that proportionate growth is also regulated by a control in the brain. Observations in the models with glucocorticoid treatment and hypothyroidism indicate that proportionate growth may have priority over catch-up growth. Thus, a failure of catch-up growth in any single dimension may result from proportionate growth regulation.

That other factors may limit catch-up growth is suggested by studies of cartilage morphology, histochemistry and *in vitro* metabolism. These have shown persistence of abnormalities into the recovery period for varying lengths of time depending on the experimental treatment. In some instances the alterations in structure and function correlate with complete or partial failure of catch-up growth.

Acknowledgements

This work was supported by NIH grants HD 12034 and HD 07074.

References

Box, G.E.P. & Jenkins, G.M. (1976). *Time Series Analysis: Forecasting and Control*, revised edition. Holden–Day, San Francisco.

Clemente, C.D., Yamazaki, J.N., Bennett, L.R. & McFall, R.A. (1960). Brain radiation in newborn rats and differential effects of increased age: II. Microscopic observations. *Neurology*, 10, 669–75.

Dearden, L.C. (1974). Enhanced mineralization of the tibial epiphyseal plate in the rat following propylthiouracil treatment: a histochemical, light, and electron microscopic study. *Anatomical Record*, 178, 671–89.

Dearden, L.C. & Espinoza, T. (1974). Comparison of mineralization of the tibial epiphyseal plate in immature rats following treatment with cortisone, propylthiouracil or after fasting. *Calcified Tissue Research*, 15, 93–110.

Dearden, L.C. & Mosier, H.D. (1972). Long term recovery of chondrocytes in the tibial epiphyseal plate in rats after cortisone treatment. *Clinical Orthopedics*, 87, 322–31.

Dearden, L.C. & Mosier, H.D. (1974a). Growth retardation and subsequent recovery of the rat tibia. A histochemical, light, and electron microscopic study. I. After propylthiouracil treatment. *Growth*, 38, 253–75.

Dearden, L.C. & Mosier, H.D. (1974b). Growth retardation and subsequent recovery of the rat tibia: A histochemical, light, and electron microscopic study. II. After fasting. *Growth*, 38, 277–94.

Ellis, S., Huble, J. & Simpson, M.E. (1953). Influence of hypophysectomy and growth hormone on cartilage sulfate metabolism. *Proceedings of the Society for Experimental Biology and Medicine*, **84**, 603–4.

Evans, H.M., Simpson, M.E., Mark, W. & Kebrick, E. (1943). Bioassay of the pituitary growth hormone. Width of the proximal epiphyseal cartilage of the tibia in hypophysectomized rats. *Endocrinology*, **32**, 13–6.

Groos, G., Mason, R. & Meijer, J. (1983). Electrical and pharmacological properties of the suprachiasmatic nuclei. *Federation Proceedings*, **42**, 2790–5.

Hatai, S. (1907). Effect of partial starvation followed by a return of normal diet on the growth of the body and central nervous system of albino rats. *American Journal of Physiology*, **18**, 309–20.

Herbai, G. (1971). Effect of age, sex, starvation, hypophysectomy and growth hormone from several species on the inorganic sulfate pool and on the incorporation *in vivo* of sulfate into mouse costal cartilage. *Acta Endocrinologica*, **66**, 333–51.

Huxley, J.S. (1932). *Problems of Relative Growth*. Methuen, London.

Inouye, S.-J.T. (1982). Restricted daily feeding does not entrain circadian rhythms in the rat. *Brain Research*, **232**, 194–9.

Krieger, D.T., Hanser, H. & Krey, L.C. (1977). Suprachiasmatic nucclear lesions do not abolish food-shifted circadian adrenal and temperature rhythmicity. *Science*, **197**, 398–9.

Moore-Ede, M.C. (1983). The circadian timing system in mammals: two pacemakers preside over many secondary oscillators. *Federation Proceedings*, **42**, 2902–8.

Mosier, H.D. (1969). Allometry of body weight and tail length in studies of catch-up growth in rats. *Growth*, **33**, 319–30.

Mosier, H.D. (1972). Decreased energy efficiency after cortisone induced growth arrest. *Growth*, **36**, 123–31.

Mosier, H.D. (1973). Control of catch-up in unidimensional and proportionate growth: Experimental studies and an hypothesis. In *Endocrine Aspects of Malnutrition, Kroc Foundation Symposia No.1*, ed. L.I. Gardner & P. Amacher, pp. 425–57. The Kroc Foundation, Santa Ynez, California.

Mosier, H.D., Dearden, L.C., Jansons, R.A. & Hill, R.R. (1977). Growth hormone, somatomedin and cartilage sulfation in failure of catch-up growth after propylthiouracil-induced hypothyroidism in the rat. *Endocrinology*, **100**, 1644–51.

Mosier, H.D., Dearden, L.C., Jansons, R.A. & Hill, R.R. (1978). Cartilage sulfation during catch-up growth after fasting in rats. *Endocrinology*, **102**, 386–92.

Mosier, H.D., Dearden, L.C., Jansons, R.A., Roberts, R.C. & Biggs, C.S. (1982). Disproportionate growth of organs and body weight following glucocorticoid treatment of the rat fetus. *Developmental Pharmacology and Therapeutics*, **4**, 89–105.

Mosier, H.D., Dearden, L.C., Roberts, R.C., Jansons, R.A. & Biggs, C.S. (1981). Regional differences in the effects of glucocorticoids on maturation of the fetal skeleton of the rat. *Teratology*, **23**, 15–24.

Mosier, H.D., Good, C.B., Jansons, R.A., Sondhaus, C.A., Dearden, L.C., Alpizar-S., M. & Zuniga, O.F. (1983a). The effect of neonatal head-irradiation and subsequent fasting on the mechanisms of catch-up growth. *Growth*, **47**, 13–45.

Mosier, H.D. & Jansons, R.A. (1967). Stunted growth in rats following X-irradiation of the head. *Growth*, **31**, 139–48.

Mosier, H.D. & Jansons, R.A. (1968). Pituitary content of somatotropin, gonadotropin, and thyrotropin in rats with stunted linear growth following head

X-irradiation. *Proceedings of the Society for Experimental Biology and Medicine*, **128**, 23–6.

Mosier, H.D. & Jansons, R.A. (1970). Effect of X-irradiation of selected areas of the head of the newborn rat on growth. *Radiation Research*, **43**, 92–104.

Mosier, H.D. & Jansons, R.A. (1971). Allometry of body weight and tail length after head X-irradiation in rats. *Growth*, **35**, 23–31.

Mosier, H.D. & Jansons, R.A. (1976). Growth hormone during catch-up growth and failure of catch-up growth in rats. *Encodrinology*, **98**, 214–9.

Mosier, H.D. & Jansons, R.A. (1985). Increase in pulsatile secretion of growth hormone during failure of catch-up growth following gflucocorticoid-induced growth inhibition. *Proceedings of the Society for Experimental Biology and Medicine*, **178**, 457–61.

Mosier, H.D., Jansons, R.A., Biggs, C.S., Tanner, S.M. & Dearden, .L.C. (1980). Metabolic clearance rate of growth hormone during experimental growth arrest and subsequent recovery in rats. *Endocrinology*, **107**, 744–8.

Mosier, H.D., Jansons, R.A. & Dearden, L.C. (1985a). Increased secretion of growth hormone in rats undergoing catch-up growth after fasting. *Growth*, **49**, 346–53.

Mosier, H.D., Jansons, R.A., Hill, R.R. & Dearden, L.C. (1976). Cartilage sulfation and serum somatomedin in rats during and after cortisone-induced growth arrest. *Endocrinology*, **99**, 580–9.

Mosier, H.D., Jansons, R.A., Swingle, K.R., Sondhaus, C.A., Dearden, L.C. & Halsall, L.C. (1985b). Growth hormone secretion in the stunted head-irradiated rat. *Pediatric Research*, **19**, 543–8.

Mosier, H.D., Sondhaus, C.A., Dearden, L.C., Zuniga, O.F., Jansons, R.A., Good, C.B. & Roberts, R.C. (1983b). Cartilage metabolism during growth retardation following irradiation of the head of the neonatal rat. *Proceedings of the Society for Experimental Biology and Medicine*, **172**, 99–106.

Osborne, T.B. & Mendel, L.B. (1914). The suppression of growth and the capacity to grow. *Journal of Biological Chemistry*, **18**, 95–106.

Pickard, G.E. & Turek, F.W. (1983). The suprachiasmatic nuclei: two circadian clocks? *Brain Research*, **268**, 201–10.

Prader, A. (1978). Catch-up growth. *Postgraduate Medical Journal*, **54** (Suppl.1), 133–46.

Prader, A., Tanner, J.M. & von Harnack, G.A. (1963). Catch-up growth following illness or starvation. *Journal of Pediatrics*, **62**, 646–59.

Savkovic, N.V. (1969). Effect of local irradiation of the head of 2 day old rats: Morphological and functional disorders and genetic changes in their progeny. In *Radiation Biology of the Fetal and Juvenile Mammal*, ed. M.R. Sikov & D.D. Mahlum, pp. 453–74. U.S. Atomic Energy Commission, Division of Technical Information, Publication CON–690501.

Salmon, W.D., Jr, Bower, P.H. & Thompson, E.Y. (1963). Sulfate metabolism by cartilage in rats during starvation. *Clinical Research*, **11**, 56.

Schjeide, O.A., Yamazaki, J.N., Haack, K., Ciminelli, E. & Clemente, C.D. (1966). Biochemical and morphological aspects of radiation inhibition of myelin formation. *Acta Radiologica*, **Suppl. 5**, 185–203.

Shalet, S.M. (1986). Irradiation-induced growth failure. *Clinics in Endocrinology and Metabolism*, **15**, 591–606.

Sinha, Y.N., Wilkins, J.N., Selby, F. & VanderLaan, W.P. (1973). Pituitary and serum growth hormone during undernutrition and catch-up growth in young rats. *Endocrinology*, **92**, 1768–71.

Snow, M.H.L. (1981). Growth and its control in early mammalian development. *British Medical Bulletin*, **37**, 221–6.

Tanner, J.M. (1963). The regulation of human growth. *Child Development*, **34**, 817–47.

Tanner, J.M. (1979). A note on the history of catch-up growth. *Bulletins et Mémoires de la Société d'Anthropologie de Paris, Ser. XIII*, **6**, 399–409.

Tanner, J.M. (1981). Catch-up growth in man. *British Medical Bulletin*, **37**, 233–8.

Tessler, R.H. & Salmon, W.D., Jr. (1975). Glucocorticoid inhibnition of sulfate incorporation by cartilage of normal rats. *Endocrinology*, **99**, 898–902.

Van den Brande, J.L. (1986). Catch-up growth. Possible Mechanisms. *Acta Endocrinologica*, **Suppl. 279**, 13–23.

Weiss, P. & Kavanau, J.L. (1957). A model of growth and growth control in mathematical terms. *Journal of General Physiology*, **41**, 1–47.

Williams, J.P.G. & Hughes, P.C.R. (1975). Catch-up growth in rats undernourished for different periods during the suckling period. *Growth*, **45**, 179–93.

Williams, J.P.G. & Hughes, P.C.R. (1978). Catch-up growth in the rat skull after retardation during the suckling period. *Journal of Embryology and Experimental Morphology*, **45**, 229–35.

Williams, J.P.G., Tanner, J.M. & Hughes, P.C.R. (1974). Catch-up growth in male rats after growth retardation during the suckling period. *Pediatric Research*, **8**, 149–56.

Winich, M. & Grant, P. (1968). Cellular growth in the organs of the hypopituitary dwarf mouse. *Endocrinology*, **83**, 544–7.

Winter, R.J. & Green, O.C. (1984). Irradiation induced growth hormone deficiency: Blunted growth response and accelerated skeletal maturation to growth hormone therapy. *Journal of Pediatrics*, **106**, 609–12.

Yamazaki, J.M., Bennett, L.R. & Clemente, C.D. (1962). Behavioral and histologic effects of head irradiation in newborn rats. In *Response of the Nervous System to Ionizing Radiation*, ed. T.J. Haley & R.S. Snyder, p. 57–73. Academic Press, New York.

Yamazaki, J.M., Bennett, L.R., McFall, R.A. & Clemente, C.D. (1960). Brain radiation in newborn rats and differential effects of increased age: I Clinical observations. *Neurology*, **10**, 530–6.

FRANK H. RUDDLE

Genomics and evolution of murine homeobox genes

Introduction

One thousand genes have now been mapped to the human genome. No obvious patterns of gene arrangement are as yet apparent. However, an important conclusion can be drawn from the provisional data already available. Numerous genes which exhibit coordinated expression are frequently unlinked, and in fact, may reside on different chromosomes. This is the case for the collagen gene family, for example, and many other similar examples exist in man as well as the mouse *Mus musculus*, where comparable gene mapping data exists (de la Chappelle, 1985).

The coordinated expression of unlinked genes, while not ruling out the cis regulation of tightly linked genes, does imply the existence of transacting regulatory factors. It can be assumed that such factors exist as DNA binding proteins. This assumption seems reasonable, since in lower eukaryotes as in yeast, such transregulatory systems have been clearly demonstrated.

Transregulatory systems may assume at the least two general forms: network and hierarchical. In the first, one assumes that the products of genes within a system are cross regulating according to cybernetic design. The activity of one gene is communicated to all other gene members of the set, ensuring overall coordination. In the network system, one imagines that the gene members encode both regulatory and functional (in the sense of enzymatic or structural) gene products. In the second or hierarchical system, one imagines a class of controller gene or genes that regulate a second set of 'functional' genes. The controller gene(s) would thus orchestrate the expression of genes that operate at some functional level. In more complicated situations, different hierarchical levels of control might exist, and one could postulate a cascade of regulatory events initiating at higher levels of the control hierarchy and influencing events at lower levels. It must be admitted that the division of regulatory systems into network and hierarchical models is artificial, but serves the useful purpose of thinking about genetic regulation in abstract terms. In reality, one imagines that elements of the network and hierarchical model intermingle.

In the mammals, there is as yet no strong, direct evidence for either the network or hierarchical models of gene regulation. In *Drosophila*, the homeotic genes provide a compelling model for a hierarchical model of gene regulation that plays an important

48   FRANK H. RUDDLE

role in the genetic control of segmentation and morphogenesis. This is particularly well established in the case of the Bithorax and the Antennapedia complexes (Gehring, 1987). In recent years, evidence has accumulated that homologous systems also exist in the higher mammals. The purpose of this article is to discuss this evidence, and evaluate an argument for a hierarchical genetic control system in mammals that serves to regulate aspects of growth, differentiation, and morphogenesis.

The homeo box
   In 1982, two groups independently demonstrated the existence of conserved domains in several of the segmentation and homeobox genes of *Drosophila* (McGinnis *et al.*, 1984; Scott & Weiner, 1984). These were termed homeo(tic) boxes. The homeobox domains are approximately 180 nucleotide bp in length and may encode a corresponding 60 amino acid domain in specific gene products. Genes possessing homeoboxes to date also invariably possess a developmental function. Such genes may be designated generically as homeobox genes.
   The homeobox domains evince interesting properties which may relate to their evolutionary origins and function. Firstly, they are highly conserved. In *Drosophila*, there are approximately seven homeobox genes designated as the Antennapedia group, because of a high level of similarity to the homeobox located in the Antennapedia gene. The degree of similarity within this group at the base pair level averages approximately 70% whereas the deduced amino acid similarity is significantly higher at 80%. It is remarkable that this degree of similarity exists in comparable homeobox gene families in numerous invertebrate and vertebrate species, including man and the laboratory mouse. Moreover, the similarities are preserved between these distantly related groups. The conservation of the Antennapedia family between the insecta and the mammalia is particularly impressive, considering their divergence over 500 million years ago. The high level of conservation implies a conserved, essential biological function.
   Other homeobox-like families have also been reported which show greater divergence in comparison to the Antennapedia group. The Engrailed group is present both in insects and mammals (Joyner *et al.*, 1985). The degree of similarity within this group exceeds that within the Antennapedia group and also extends over a larger domain. However, the similarity between the A group and the En group is only 50% at the DNA bp level. More divergent families of homeobox domains have been reported in *Drosophila*, such as those characterized by Bicoid and Paired (Bopp *et al.*, 1986). Their counterparts in mammals have not yet been reported.
   Sequence comparisons of the homeobox domain show that a subdomain shows a homology with domains of the yeast MAT genes that serve a DNA binding function (Shephard *et al.*, 1984). This subdomain region extends over approximately 20 bp, and encodes a helix turn helix DNA binding motif at the deduced amino acid sequence

level. This homeobox domain also shows a weak relatedness to the helix turn helix DNA binding domains of prokaryotic DNA binding proteins, such as lac and lambda repressors (Laughon & Scott, 1986). It is of interest that the helix turn helix domain in the case of the A homeobox family shows a high level of conservation, suggesting a functional role in DNA binding. This possibilty has been strengthened by direct experimentation which provides evidence for a DNA binding capacity of homeobox proteins in both *Drosophila* and in the mouse. All these data taken together suggest that homeobox proteins may function at least in part as DNA binding proteins. Moreover, there is emerging evidence implicating the homeobox proteins as having regulator properties in the sense of transacting control.

Homeobox genomics

A series of reports have established the map positions of the murine homeobox genes (Ruddle *et al.*, 1986). Corresponding mapping studies have also been carried out for the human species (Ruddle *et al.*, 1986). In the mouse, four loci have been identified for the Antennapedia type homeobox genes. These have been named Hox–1 to –4 and are located on chromosomes 6, 11, 15 and 12 respectively (Figure 1). Two engrailed type homeobox genes have also been mapped in the mouse (En 1 and 2), and they have been assigned to chromosomes 1 and 5. In this discussion, I shall restrict further mention to the A type genes. The homeobox genes exist primarily as gene clusters. Hox–1 has a minimum of six genes numbered Hox–1.1 to –1.6 (Figure 1). This cluster has been established by the analysis of overlapping cosmids and extends over a distance of approximately 70–80 kb. Hox–2 is also a cluster, consisting of five genes, Hox–2.1 to –2.5. Hox–3 has one well established gene (Hox–3.1), but a second gene has been alluded to by Gruss. Hox–4 has but a single gene to date (Lonai *et al.*, 1987).

The subchromosomal location of the homeobox gene loci has also been established, both at low resolution by somatic cell genetics (Hart *et al.*, 1985) and by *in situ* (Rabin *et al.*, 1986) methods, and at intermediate resolution by Mendelian (Hart *et al.*, 1987) techniques. These studies in the mouse have established that Hox–1 is less than 1 cM distant to *Hd* (hypodactyly), a morphogenetic mutant that affects skeletal development; Hox–2 is about 1 cM distant from rex (*Re*), a mutation that affects hair development; and Hox–3 is approximately 1 cM distant from velvet (*Ve*), a mutation that also affects hair development, and in the homozygous condition affects neural plate development (Green, 1981). Hox–4 has not yet been mapped to a subchromosomal location on chromosome 12.

Human homeobox loci cognate to the mouse have also been mapped. The cognate relationships are shown in Table 1. The Hox–1 cognate has been mapped to the short arm of chromosome 7; Hox–2 to the long arm of chromosome 17; and Hox–3 to human 12q. There are indications that the human loci are also composed of multiple genes, but their organization has not been determined in a degree of detail comparable

50   FRANK H. RUDDLE

Figure 1. Murine homeobox gene map positions.

The decimal designations are the names for the homeobox genes and their homeoboxes. The designations cited under the homeobox genes are older, trivial names. The numbers in decade intervals estimate distance in kilobase units. En designates the engrailed family. Hox represents the A family.

The Murine Homeo Box Gene System

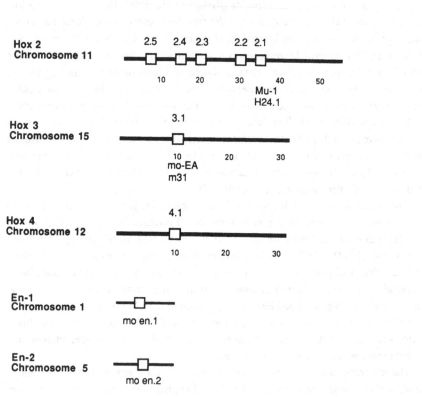

Table 1. *Human cognates of murine homeobox gene loci. Cognate relationships between the human and mouse homeobox gene loci and their chromosome map positions. Dashed lines indicate situations in which cognate relationships have not yet been determined*

Locus	Mouse chromosome	Human chromosome
Hox–1 6 7p		
Hox–2	11	17q
Hox–3	15	12q
Hox–4	12	–
C8 (provisional Hox–5)	–	2q

to the mouse. A fourth homeobox gene locus (C8) has been reported (Simeone *et al.*, 1987) and mapped to human chromosome 2q. There are indications that the human loci are also composed of multiple genes, but their organization has not been determined in a degree of detail comparable to the mouse. A fourth homeobox gene locus (C8) has been reported (Simeone *et al.*, 1987) and mapped to human chromosome 2q. To date, no indication of homeology to previously described mouse homeobox loci has been determined for this gene (homeologous is used to designate genes which show an homologous relationship within a species). Similarly, a human cognate for murine Hox–4 has not yet been identified.

To date, a total of 13 A type homeobox genes have been mapped with certainty in the mouse. The mapping of an additional two genes (Hox–1.7, and –3.2) have been informally reported for a possible total of 15. Southern blotting experiments using Antennapedia probe provides estimates of 18–20 gene copies, but these may be underestimations, since more divergent homeobox sequences may not register a signal with this probe. It is likely that additional genes exist which have not yet been mapped.

It is of interest that genes sharing an evolutionary relatedness are found on human chromosomes 7 and 17. The homeologous genes are Hox–1 and –2, neural polypeptide (NPY) and pancreatic polypeptide (PPY), epidermal growth factor–1 (ERB–B1) and epidermal growth factor–2 (ERB–B2), collagen type I, alpha 2 (ColIA2) and collagen type I, A1 (ColIA1) on chromosomes 7 and 17, respectively (Figure 2). The relationship is all the more striking because of similarities between the Hox–1 and –2 loci involving Hox–1.1, –1.2, and –1.3 and Hox–2.1, –2.2, and –2.3. We find that greater than 90% identity exists between Hox–1.1 and –2.3, –1.2 and –2.2, –1.3 and –2.1 at the deduced amino acid level (Figure 3). The Hox boxes –2.1 and –1.3 show 100% identity! Moreover, the spacing between these genes match well when they are ordered –1.1, –1.2, –1.3 for Hox–1 and –2.3, –2.2, and

–2.1 for Hox–2 (Hart & Ruddle, 1987). This ordering also brings the similar homeoboxes between the loci into correspondence on the basis of sequence similarity (Figure 3). Moreover, in this order the direction of transcription for both loci is the same (Figure 3). These similarities between the two Hox loci 1 and 2, together with

Figure 2. Homeologous relationship between Hox–1 and Hox–2 genes and linked genes.

Hox–1 maps to mouse chromosome 6 and Hox–2 to mouse chromosome 11. The order of the genes in the Figure is not meant to signify linkage relationships in the corresponding chromosomes.

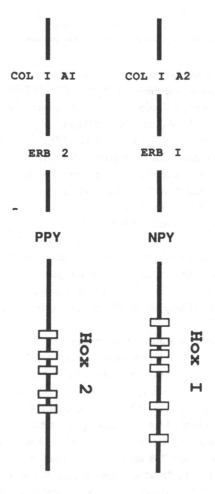

three linked homeologous loci, suggest that chromosome 7 and 17 may have arisen from a common chromosome element, possibly as the result of a duplication event.

When we examine the murine homeobox genomic relationship, a second possible internal homology presents itself. Hox–2 on chromosome 11 shows a close linkage relationship to red (*Re*), a mutation that affects hair formation. Other hair formation loci are also closely linked to *Re*, namely bareskin (*Bsk*), alopecia (*Al*), and lustrous (*lt*). Hox–3 on chromosome 15 is closely linked to velvet (*Ve*), another hair texture locus, and to other hair formation loci such as caracul (*Ca*), shaven (*Sha*), and naked (*N*) (Green, 1981). A recent finding of interest is that genes encoding cytokerations also map to these two regions on chromosomes 11 and 15 (Nadeau, personal communication). Although not proven, one might hypothesize that the hair formation loci may be identical to loci that encode individual keratins. This is an appealing notion, since it would provide an explanation for the more severe morphological conditions that certain of the hair formation loci present when homozygous. For

Figure 3. Homeologous relationships between murine homeobox genes.

Homeobox genes are indicated by decimal notations. Homeologies are indicated by bars with double arrows. Homologies in terms of amino sequence similarities are indicated by bars to *Drosophila* homeoboxes in the deformed (Dfd), Antennapedia (Antp), and interabdominal 7 (Iab7) genes. Anterolimits of expression are indicated by indicated regions of the central nervous system. Dotted lines with arrows indicate the direction of transcription.

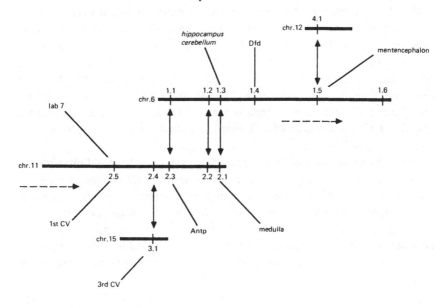

example, in the homozygous state, velvet shows an arrest of neural plate formation
and failure of neurulation. It would be satisfying to explain both the hair variant and
the ectodermal failure by a common defect in a particular cytokeratin gene. Further,
molecular genetic analyses of these loci will prove or disprove this hypothesis, as
well as the possible internal homology (homeology) between the murine Hox–2 and
Hox–3 regions. However, in terms of the internal homology argument, it should be
pointed out that Hox–2.4 shows a strong sequence similarity to Hox–3.1. If the
internal homology argument is correct, we might also predict a similarity between
Hox–2.5 and Hox–3.2. If there is evidence for homeology between Hox–2 and
Hox–3 in the mouse, we would also expect a similar relation in the human genome.
Unfortunately, human Hox–3.1 and –2.4 have not yet been sequenced.

Homeobox gene expression in the central nervous system (CNS)

A number of gene expression studies have been reported largely in man and
mouse for the A type homeobox genes (see review, Fienberg et al., 1987). In this
section, I shall concentrate on studies dealing with the CNS and review those reports
dealing primarily with the mouse that describes expression by Northern analysis or
by in situ methods in late fetuses of around 13–14 days of gestation. No attempt will
be made to follow the expression of individual transcripts. Patterns of expression of
homeobox genes in the CNS ascertained by studies in our own laboratory are
summarized in Figure 4.

Hox–1 transcripts have been shown to be expressed in both the brain and the
anterior spinal cord. Hox–1.5 shows expression in the entire myelencephalon
extending posteriorly in the form of a gradient into the spinal cord. Hox–1.3
(Odenwald et al., 1987) has been detected in the cerebellum, hippocampus, and
spinal cord. The other homeobox genes in Hox–1 have been insufficiently studied to
determine their time and place of expression in the CNS. However, based on
Hox–1.5 and –1.3, the posterior brain and anterior spinal cord represent the anterior
boundaries of transcript expression. Transcripts also are expressed in more posterior
levels of the CNS.

Hox–2 expression is centred in more posterior levels of the brain and spinal cord in
comparison to Hox–1, although one must keep in mind the fragmentary character of
the data. Hox–2.1 shows an anterior boundary of expression in the prospective
medulla (posterior myelencephalon) extending into the spinal cord. Hox–2.5 is
expressed anteriorly at the level of the first cervical vertebra, and more weakly
posteriorly in the spinal cord. Hox–2.2 to –2.4 are expressed in 13.5 day embryos,
but further work is required to assay their positions of expression in the CNS and/or
other organs.

Hox–3.1 expression has been thoroughly studied in the CNS in 13.5 day,
newborn, and young adults (Awgulewitsch et al., 1986; Utset et al., 1987). The

anterior level of expression in the CNS is at the level of the third cervical vertebra. Expression is highest at the anterior limit and extends posteriorly at reduced levels as a gradient.

Hox–4.0 has not been studied in the CNS systematically. However, it is of interest that Hox–4.0 has a high sequence similarity with the Hox–1.5 homeobox (Figure 3).

Evolution of the homeobox A gene system

The evolution of the A type homeobox gene family will become clear when extensive DNA sequence information is available. Currently, not much can be said, since only homeobox sequences are available. However, it may be useful to examine this information together with the other genomic and expression data in order to suggest evolutionary models which can be used to guide more detailed sequencing activities in the future.

A starting point is the realization that the Antennapedia sequence is very similar to the consensus sequence for the so called A type gene family in *Drosophila*, mouse, and man (Hart & Ruddle, 1987). This result may be biased in the mammals, since Antennapedia was used as the original probe in the identification of this family. However, this was not the case in *Drosophila* where the Antennapedia, Bithorax, and related genes were defined in terms of their map positions and genetic functions as determined by mutations. Thus, if one accepts the possibility that this gene family has

Figure 4. Expression of homeobox gene transcripts in the mouse 13.5 day fetal central nervous system.

Stippled regions indicate transcript expression on the basis of *in situ* hybridization.

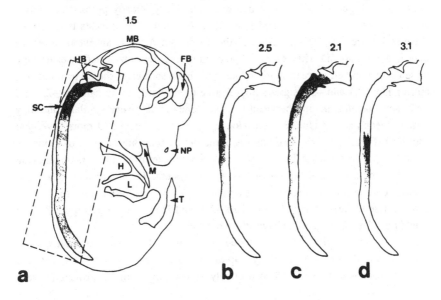

arisen through mechanisms such as gene duplication and divergence, then one may consider Antennapedia to represent an ancestral type, and its map position to be 'primitive'. Below, I shall discuss the homeobox genes as units within epigenetic arrays. One might think of the more primitive units becoming rapidly fixed in terms of their structure and sequence, since a high degree of interdependency would arise between the homeobox genes and other genetic elements and programs in which they operate. New genes arising by duplication would diverge, then likewise become fixed as a consequence of their essential epigenetic interactions. If this model of a divergence, fixation cycle is correct, then one should be in a position to trace the evolution of the homeobox genes readily. A possible pitfall in the use of the homeobox sequence as a delineation of evolutionary relationship is their general high level of identity and the possibility that chance variation or drift may introduce false similarities and differences. These problems are minimized by using additional features, such as the specific patterns of deduced amino acid substitutions and features in other parts of the gene. A final conclusion will be possible, as stated above, when extensive sequence information is available in a number of representative species.

On the basis of the arguments above and taking into account the reservations advanced, the following model is proposed for the evolution of the vertebrate homeobox system (see Figure 3). Hox–2.3 because of its sequence similarity with the Antennapedia homeobox is regarded as primitive. There is a strong similarity between Hox–2.3 and –2.2 (89 bp/92 AA) and between Hox–2.2 and –2.1 (80/85). One might surmise that Hox–2.3, –2.2, and –2.1 are descendants of an ancestral group A, B and C, and that this ancestral series arose by gene duplication of the kind indicated in Figure 5A. The high degree of similarity already pointed out above between Hox–1.1, –1.2, –1.3 and Hox –2.3, –2.2 and –2.1 suggests that both of these series arose by a duplication of the postulated A, B, C ancestral series, as indicated in Figure 5D. This is a useful prediction, since it can be tested by examining these genes in other, more distantly related vertebrate species. It should be noted that murine Hox–2.1 and –2.2 show a high degree of similarity to human Hox–2.1 and –2.2, and that the spacing between the two genes is also very similar. One may further speculate that Hox–2.4 and Hox–1.5 are more recent and more divergent additions (Figure 3). This prediction is borne out by their high degree of sequence divergence. It is of interest that the major A type loci in terms of numbers of genes are related to the postulated origins of this A type system. The minor loci located on separate chromosomes and containing only one or a few genes are related to the most divergent sequences, names Hox–3.1 to –2.4, and Hox–4 to –1.5 (Figure 3). We would postulate these genes to have arisen most recently.

In this regard in the mouse, as stated above, it is of interest that coat genetic markers that affect hair morphology map to positions adjacent to Hox–2 and Hox–3 loci on chromosomes 11 and 15, respectively. If homology could be proven for these genes, then a strong case for a homeologous relationship could be made between portions of chromosomes 11 and 15. If more divergent genes have indeed arisen more recently, then this allows an additional prediction that the homologs of these genes may be altogether absent in more primitive forms of the deuterostome series and have no counterpart among the protostomes. This may possibly be the case in respect to Hox–2.4 and –1.5 and *Drosophila* genes. However, it should also be noted in agreement with the model proposed here that the less divergent mouse gene Hox–1.4 has a relationship to the *Drosophila* homeobox gene *Dfd*, and that Hox–2.5 shows a weaker resemblance to Iab7. These relationships suggest that the ancestral organization of A type homeobox systems especially as it pertains to the major loci and possibly minor loci was already in place prior to the divergence of the protostome and deuterostomes.

Genes syntenic to the Hox genes may play a role in growth and differentiation

If the assumption that Hox loci are controller gene complexes, then one can further postulate that the Hox genes regulate 'target' genes at more distant sites in the genome. However, since transregulation must depend on conserved DNA binding domains or operator sequences, and since these operators most probably arose within the regulatory domains of the Hox genes themselves, then it follows as a possibility that target genes may have been dispersed during evolution, but others as a consequence of chance or of adaptiveness may have retained a proximal position. For these reasons, it is of interest to examine the genes that map to positions syntenic to the homeobox loci. The Hox–2 linkage unit on human chromosome 17 yields useful information, since many genes have been mapped to this chromosome. Human chromosome 17 also appears to be unusually conserved during primate evolution, since all of its genes showing an homology with the mouse project to a single mouse chromosome, namely chromosome 11 (see Table 2).

Examination of human chromosome 17 shows an apparent high concentration of genes that relate to either growth or differentiation. Of the 31 genes of known function, one can classify 11 as having a possible involvement in growth regulation. These are chorionic somatomammotropin hormone 1 and 2 (CSH 1 and 2), growth hormone 1 and 2 (GH 1 and 2), nerve growth factor receptor (NGFR), epidermal growth factor receptor type 2 (EGFR–2), protein kinase C, alpha gene (PKCA), pancreatic polypeptide (PPY), thymidine kinase 1 (TK1), tumor protein p53 (Tp53), and granulocyte colony stimulating factor (GCSF). Twelve genes can be classified as having a relationship to differentiation. These are collagen, type 1, alpha 1 (COL 1A1), crystallin, beta polypeptide 1 (CRY B1), glucocorticoid receptor (ERBA1), gastrin (GAS), myosin heavy chain 1–4 (MYY1–4), tissue specific extinguisher 1 (TSE1), tau (microtubule) protein (TAU), myeloperoxidase (MPO), and creatine kinase (CK). The remaining genes of possibly housekeeping, constitutive function

Table 2. *Homology Projections of Homeobox Chromosomes between Man and Mouse. Homologies have been projected from human to mouse and vice-versa. For example, human 2 bears 16 genes, the homologs of which have been mapped in the mouse. The murine genes have been mapped to chromosomes 1, 2, 6, 7, 11, 12, 17 and 19. Chromosomes in both species to which homeobox genes have been mapped are underlined*

Human	into	Mouse
2 (16 genes)		1, 2, <u>6</u>, 7, <u>11</u>, <u>12</u>, <u>17</u>, 19
7 (10)		2, 5, <u>6</u>, <u>11</u>, 13, 16
17(11)		<u>11</u>
12 (16)		4, <u>6</u>, 8, 10, <u>15</u>

Mouse	into	Human
6 (24)		<u>2</u>, 3, 4, <u>7</u>, 11, <u>12</u>, 20
11(17)		<u>2</u>, 5, <u>7</u>, 16, <u>17</u>, 22
12(9)		<u>2</u>, 14
15(14)		8, <u>12</u>, 16, 22

number 8 of 31. Some of these could conceivably be assigned to the growth or differentiation categories. They are glucosidase alpha acid (GAA), galactokinase (GALK), glucose–6–phosphate dehydrogenase–like (G6PDL), peptidase E (PEPE), placental alkaliane phosphatase-like (Regan isozyme, PLAPL), ribosomal protein L25 (RPL25), ribosomal protein S17B (RPS17B), and uridine 5'–monophosphate phosphohydrolase 2 (UMPH2). For comparison, human chromosome 1 was examined. Fifty-eight genes performing discrete functions were identified. Thirty-one could be assigned to housekeeping functions, 17 to specialized (growth and differentiation) functions, and ten could not be readily assigned. If we assume that the ratio of specialized to housekeeping genes is 0.5, then the number of specialized genes on chromosome 17 is weakly significant (χ^2 test). Contrariwise, if we assume the ratio of specialized to housekeeping is similar to that seen for chromosome 1, namely 0.35, then the number of specialized functions on chromosome 17 is highly significant. It is of interest that chromosome 7 (Hox–1) also shows an enrichment of specialized gene function in a manner similar to chromosome 17. Chromosome 2 (Hox–4) has roughly an equal number of specialized to housekeeping genes, whereas, chromosome 12 (Hox–3) is enriched for housekeeping genes to a degree similar to chromosome 1. One must keep in mind the hazards of such an analysis. Few genes have been mapped to the chromosomes, the classification into specialized and housekeeping genes is fraught with uncertainty and vague at best. However, keeping these limitations in mind, one is still left with the impression that the chromosomes bearing the major A-type homeobox loci Hox–1 and –2 have an enrichment for specialized gene functions, whereas the chromosomes bearing the minor A-type loci show little or no enrichment.

Genomic mechanisms involved in homeobox gene family linkage relationships in man and mouse

In man, there is compelling evidence for a homeologous relationship between Hox–1 and –2 on chromosomes 7 and 17, respectively. The same is true of Hox–2 and –3 in the mouse, which correspond to human chromosomes 17 and 12. There is also weaker evidence supporting a homeologous relationship between Hox–4 and Hox–1. However, the homeologous relationships do not exist in all combinations. That is to say, on the basis of homeobox sequences there is no close relationship between Hox–1 and –3, –2 and –4, or –4 and –3. This argues against a tandem series of two chromosome duplication events. Moreover, if one examines linked markers, one observes similar patterns of homeology. How might these patterns have arisen? Figure 5 suggests one possible model consistent with the currently observed linkage relationships.

Consider first that the homeobox gene family has arisen by a series of tandem gene duplications, such that there are approximately seven genes in tandem array with flanking genes BC proximal and AD distal (Figure 5A). Following a chromosome segment duplication, two identical strands are generated (Fig. 5B). This is followed by an inversion in one strand around the homeobox gene region between the proximal and distal flanking genes (Fig. 5C). A third event involves multiple translocations within the homeobox gene region and other chromosomes (Fig. 5D). This is the most unlikely step, since it implies either two independent serial events within the homeobox gene complex, or two simultaneous events within this region involving four chromosomes. The latter is favored since complicated events of this type have been documented. Finally, the chromosomes are rearranged to emphasize their homeologous relationships and assigned arbitrary designations (Fig. 5E). The strands now conform to the observed linkage and homeologous relationships in the human and mouse homeobox gene systems. The properties of the system are as follows: (1) the core chromosomes alpha and beta show an homology between homeobox genes 3–5 and 3'–5', and the distal flanking markers A and A'. These linkage groups would correspond to Hox–1 and –2. The peripheral chromosomes gamma and delta show a distant relatedness with respect to their homeobox genes and no relationship with regard to proximal flanking genes. There is an homeologous relationship with respect to the distal flanking genes. Each core chromosome shows a close homeologous relationship with a specific peripheral chromosome as in the case of alpha–gamma and beta–delta. This is best exemplified by the murine Hox–2 and –3.

All other combinations show weak relationships at best. This scheme is consistent with the existing observed linkage patterns in man and mouse, where alpha, beta, gamma, and delta correspond to Hox–1, –2, –4, and –3. The relationships involving alpha and gamma are at least well established because of the uncertainty regarding the mouse/human homology for the gamma linkage group and its contained homeobox. It

Figure 5. Model for the evolutionary genomics of the homeobox genes in man and mouse.

The numerals indicate homeobox genes which have arisen by amplification of the unequal crossing over type. B and C represent proximal flanking genes, and A and D distal genes. A. The original condition showing the ancestral linkage arrangement. B. Duplication following a polyploidy or aneuploidy event. C. Inversion in one segment between proximal and distal flanking gene markers. D. Translocation within the homeobox cluster transfers segments to other autosomes. E. Rearrangement of the segments to show homeologies and to correspond more closely to currently observed linkage. Greek symbols are arbitrary designations to facilitate discussion in text.

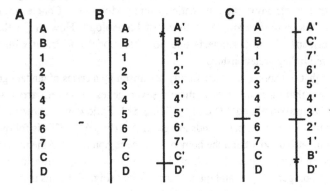

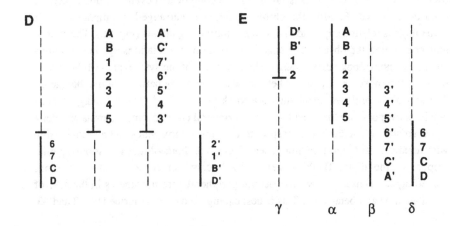

is also likely that A-type homeobox genes will be found that map to other chromosomes as a consequence of additional genome rearrangements. Nevertheless, the current model is consistent with available information and provides a series of hypotheses that can be readily tested.

Possible genomic relationships between the homeobox gene systems in *Drosophila*, man and mouse

The A-type homeobox sequences between *Drosophila* and mammals are remarkably similar, and indicate an evolutionary relationship. Additional similarities involve the assemblage of the genes to clusters and functional parallels, such as the overlapping patterns of expression in the central nervous system already remarked upon above, as well as the existence of DNA binding sites in the 5' flanking regions at which the same gene products interact. However, differences also exist. The number of A-type homeobox genes appears to be significantly less in *Drosophila* than the number detected in the mammals, roughly by a factor of two. The *Drosophila* genes are also considerably larger in size by a factor of 5- to 8-fold. It is interesting to speculate on the evolutionary bases for these similarities and differences. A possible explanation is shown in Figure 6. We assume the homeobox complex has arisen by gene duplication in precambrian forms prior to the divergence of the proro- and deuterosomia. Certain precambrian forms were in many respects highly developed in terms of their metameric organization, as well as their degree of cephalization and caudalization. If the homeobox genes are involved in these attributes, a case can therefore be made for their relatively advanced organization as a fairly large complex of genes as shown in Figure 6B. As argued later below, as the array of genes increased by duplication, the individual genes may have acquired specificity with respect to regions of the anatomy affected, such that genes at one end of the array affected anterior structures, whereas those at the opposite end controlled posterior pattern formations. These ideas, especially as they pertain to the heterotic adaptiveness of homeotic genes in *Drosophila*, have been advanced originally by Lewis (1985) in his pioneering studies on the subject. It is possible that the more anterior and posterior domains have become separated by chromosomal rearrangement to give rise to the currently recognized Antennapedia and Bithorax complexes (see Figure 6C). Although not necessarily required, the transcription direction of the genes may be oriented in the same direction. It is interesting that this is largely so for the A–C and the B–C, both in a posterior to anterior direction in a way consistent with the scheme shown in Figure 6C.

A fundamental difference between the *Drosophila* and mammalian systems may reside in the proposed major duplication event indicated in Figure 6D, and followed by translocations as shown in Figure 6E (see also Figure 5). The duplication would be consistent with the currently observed ratio of A-type homeobox genes detected in

Figure 6. Model for the evolution of A-type homeobox gene complexes in *Drosophila* and mammals.
A. Origin of homeobox complexes by gene duplication prior to the divergence of protosomes and deuterostomes. *B*. Early homeobox gene array organized in such a manner that there is a relationship between gene order and the anterior–posterior anatomial axis; it is assumed that genes 1, 2, 3, ... control caudal pattern formation, whereas genes 8, 7, 6, ... operate on anterior structures; it is also assumed that translation proceeds in a 1 to 9 direction. *C*. Bithorax and Antennapedia complexes are separated by chromosome rearrangements. *D*. In deutrosomic evolution, the gene complex is duplicated. *E*. The contemporary condition is generated by inversion and translocation mechanisms. In a comparison of C and E, the overall similarities of polarity and function are retained, but gene number differs. Note similarities between E and Figure 3.

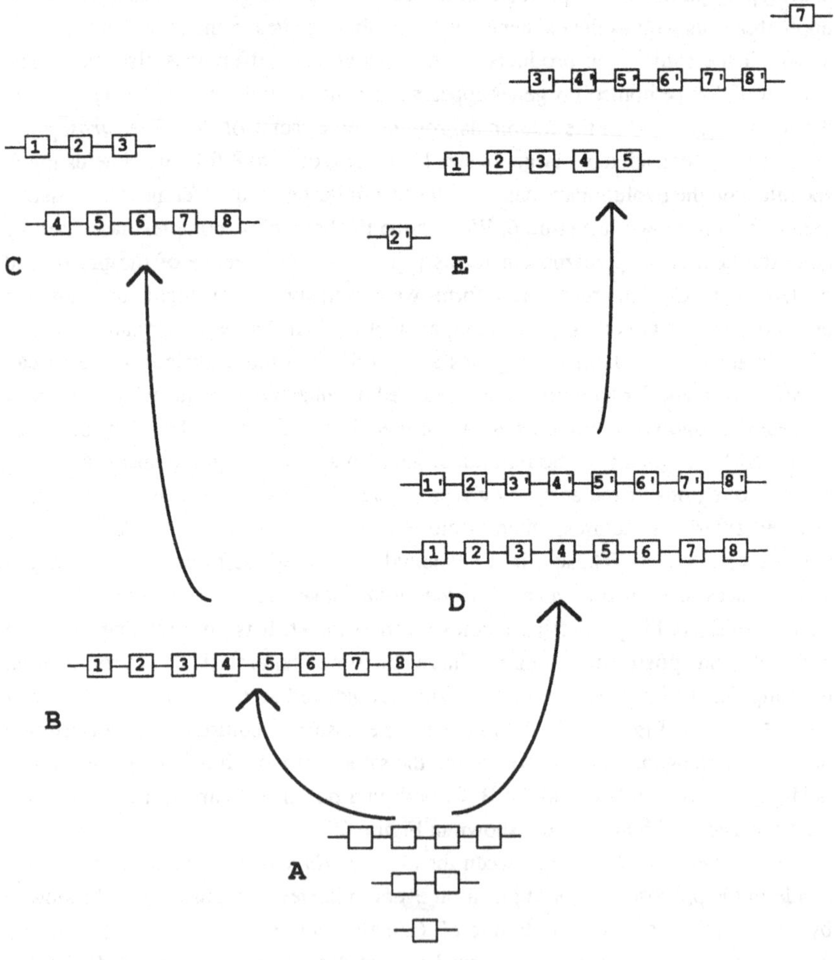

Drosophila versus the mammalian forms. If the duplicated genes were retained in the evolution of the deuterostomia, it is possible that certain attributes of affinity with *Drosophila* might still be maintained, as for example, the overall relationship of the array in respect to the anterior–posterior axis, the uniformity of transcription direction in posterior to anterior direction, and a general parallel relationship between A type homeobox sequences in *Drosophila* and mammals. Although not perfect, there appears to be general consistency in the genomic and functional relationships as depicted in Figures 3 and 6. Finally, an additional speculation is advanced concerning the size differences between the *Drosophila* and mammalian A type homeobox genes. The larger *Drosophila* genes whose increase in restricted to cis regulatory flanks may represent an adaptation to a smaller overall number of genes. The deuterostome solution involving a larger gene set may require a substantially smaller control domain for each individual gene. Albeit tentative and admittedly highly speculative, I believe it is worthwhile to advance these hypotheses as a means to guide our thoghts on controller gene systems. The hypotheses have the real and ultimate virtue of being susceptible to critical evaluation by experimentation.

Cyber-genetic role of the homeobox gene complex

The postulated evolution and function role of the homeobox gene complex is presented in Figure 7. The ancestral gene is postulated to produce a product possessing a homeobox DNA binding domain that interacts with a cis regulatory element of the gene, thus conferring an autoregulatory function. Direct evidence for 5' DNA binding sites (operator sites) have been reported in both *Drosophila* (Desplan *et al.*, 1985) and the mouse (Fainsod *et al.*, 1986) (Figure 7A; circuit alpha). The cis regulatory domains are also regulated by other transcriptional regulatory factors, transducing environmental information (Figure 7A; beta factors). The gene product is presumed to also influence the expression of genes at a lower level of hierarchy (i.e. target genes not shown) or to provide other epigenetic control functions (Figure 7A; gamma circuit). Duplication of the gene preserves the original features, but provides new delta circuits (Figure 7B) which serve to cross-regulate the new genes. Recently reported experiments provide direct evidence for the transregulation of homeobox genes. Subsequent changes in the product or the cis regulatory domains now permit the two genes to acquire a switching function. Additional elements as shown in Figure 7C provide additional complexity. The elaborate arrays can be expected to have design features such that complex environmental beta signals can be detected and transduced into complex patterns of gamma control signals, which in turn can coordinate the expression of target genes crucial to growth and differentiation. An array of six homeobox genes could readily accommodate billions of informational states. It will be of interest in the years ahead to test the validity of this model and to explore all of its ramifications.

Figure 7. Evolution of cyber-genetic controller genes.

A. Single gene with autoregulatory alpha circuit mediated by a DNA binding protein interacting with a specific operator site in the 5' cis regulatory domain, environmental β signals in the form of transregulators, and control gamma signals reulating target gene(s) presumed involved in growth and differentiation functions. *B.* Duplication of the genes giving rise to a switching machine. *C.* A more elaborate controller gene complex capable of transducing environmental information into control instructions, thus mediating the coordinated response of target genes not shown.

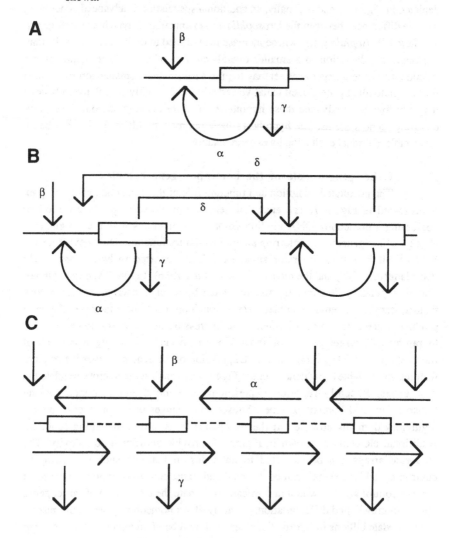

References

Awgulewitsch, A., Utset, M.F., Hart, C.P., McGinnis, W. & Ruddle, F.H. (1986). Spatial restriction in expression of a mouse homeo box locus within the central nervous system. *Nature*, **320**, 328–35.

Bopp, D., Burri, M., Baumgartner, S., Frigerio, G. & Noll, M. (1986). Conservation of a large protein domain in the segmentation gene *paired* and infunctionally related genes of *Drosophila. Cell*, **47**, 1033–40.

de la Chapelle, A. (ed.) (1985). *Human Gene Mapping 8*. Karger, Basel.

Desplan, C., Theis, J. & O'Farrell, P. (1985). The *Drosophila* developmental gene, engrailed, encodes a sequence-specific DNA binding activity. *Nature*, **318**, 630–5.

Fainsod, A., Bogarad, L.D., Ruusala, T., Lubin, M., Crothers, D.M. & Ruddle, F.H. (1986). The homeo domain of a murine protein binds 5' to its own homeo box. *Proceedings of the National Academy of Sciences, USA*, **83**, 9532–6.

Fainsod, A., Awgulewitsch, A. & Ruddle, F.H. (1987). Expression of the murine homeo box gene Hox–1.5 during embryogenesis. *Developmental Biology*, in press.

Fienberg, A.A., Utset, M.F., Bogarad, L.D., Hart, C.P., Awgulewitsch, A., Ferguson-Smith, A., Fainsod, A., Rabin, M. & Ruddle, F.H. (1987). In *Recent Advances in Mammalian Development*. In press. Academic Press, Fl.

Gehring, W.J. (1987). Homeotic genes, the homeo box, and the spatial organization of the embryo. *Harvey Lectures, Series 81*, pp. 153–72.

Green, M.C. (ed). (1981). Catalog of mutant genes and polymorphic loci. In *Genetic Variants and Strains of the Laboratory Mouse*, pp. 8–278. Gustav Fischer Verlag, Stuttgart.

Hart, C.P., Awgulewitsch, A., Fainsod, A., McGinnis, W. & Ruddle, F.H. (1985). Homeo box gene complex on mouse chromosome 11: molecular cloning, expression in embryogenesis, and homology to a human homeo box locus. *Cell*, **43**, 9–18.

Hart, C.P., Dalton, D.K., Nichols, L., Hunihan, L., Roderick, T.H., Langley, S.H., Taylor, B.A. & Ruddle, F.H. The Hox–2 homeo box gene complex on mouse chromosome 11 is closely linked to *Re. Genetics*, in press.

Hart, C.P., Fainsod, A. & Ruddle, F.H. Sequence analysis of the murine Hox–2.2, –2.3, and –2.4 homeo boxes: evolutionary and structural comparisons. *Genomics*, in press.

Joyner, A.L., Lebo, R.V., Kan, R., Tjian, R., Cox, D.R. & Martin, G.R. (1985). Comparative chromosome mapping of a conserved homeo box region in mouse and human. *Nature*, **314**, 173–5.

Laughton, A. & Scott, M.P. (1984). Sequence of *Drosophila* segmentation gene: protein structure homology with DNA binding proteins. *Nature*, **310**, 25–31.

Lewis, E.B. (1985). Regulation of the genes of the Bithorax complex in *Drosophila. Cold Spring Harbor Symposium on Quantitative Biology*, **50**, 155–64.

Lonai, P., Arman, E., Czosnek, H., Ruddle, F.H. & Blatt, C. New murine homeoboxes: structure, chromosomal assignment, and differential expression in adult erythropoiesis. *DNA*, in press.

McGinnis, W., Garber, R.L., Wirz, J., Kuroiwa, A. & Gehring, W.H. (1984). A homologous protein-coding sequence in *Drosophila* homeotic genes and its conservation in other metazoans. *Cell*, **37**, 403–9.

Odenwald, W.F., Taylor, C.F., Palmer–Hill, F.J., Friedrich, Jr., V., Tani, M. & Lazzarini, R.A. Expression of a homeo domain protein in non-contact inhibited cultured cells and post-mitotic neurons. *Genes & Development*, in press.

Rabin, M., Ferguson-Smith, A., Hart, C.P. & Ruddle, F.H. (1986). Homeobox loci mapped in homologous regions of human and mouse chromosomes. *Proceedings of the National Academy of Sciences, USA*, **86**, 9104–8.

Ruddle, F.H., Hart, C.P., Rabin, M., Ferguson-Smith, A. & Pravtcheva, D. (1986). Comparative genetic analysis of mammalian homeo box genes. In *Human Genetics, Proceedings of the 7th Intl. Congress, Berlin*, ed. F. Vogel. Springer–Verlag, in press.

Scott, M.P. & Weiner, A.J. (1984). Structural relationships amaong genes that control development: sequence homology between the Antennapedia, ultrabithorax, and fushi tarazu loci of *Drosophila. Proceedings of the National Academy of Sciences, USA*, **81**, 4115–9.

Shepherd, J.C.W., McGinnis, W., Carrasco, A.E., DeRobertis, E.M. & Gehring, J. (1984). Fly and frog homeo domains show homologies with yeast mating type regulatory proteins. *Nature*, **310**, 70–1.

Utset, M.F., Awgulewitsch, A., Ruddle, F.H. & McGinnis, W. (1987). Region specific expression of two mouse homeo box genes. *Science*, **235**, 1379–82.

Note added in press: This article was prepared two years prior to publication and there was no opportunity to revise the text during that period. Therefore, many ideas and interpretations have changed, and in some instances, radically. For example, the notion of chromosomal duplication followed by chromosomal translocation to account for the distribution of homeobox gene clusters is better explained on the basis of more complete data, simply by two serial chromosome (or genomic) duplications. The terminology has also changed in some instances, and where the text reads "homeologous", one should substitute "paralogous". More modern and detailed expositions of these ideas will be published in the *Proceedings of the National Academy of Sciences, USA* in 1989 by the author, F.H. Ruddle.

Part II

Nutrition, growth and body composition

R.G. WHITEHEAD, A.A. PAUL &
E.A. AHMED

United Kingdom Department of Health and Social Services 'Present-day infant feeding practice' and its influence on infant growth

The dietary advice set out by the Department of Health and Social Services in *Present-day Practice in Infant Feeding* differed substantially from the practices which were current when the booklet was produced, as well as from those which were common when the NCHS (Hamill, 1977) and Tanner, Whitehouse & Takaishi (1966) growth reference standards were being compiled. This short review assesses the likely consequences for infant growth if the DHSS advice were more widely adopted. We have been investigating prospectively derived growth patterns in the city of Cambridge, where a much higher proportion than usual of mothers have changed their infants' feeding practices in accordance with official recommendations.

Feeding practice

The basic recommendation in DHSS (1980), recently updated in a statement from the Committee on the Medical Aspects of Food Policy (COMA, 1987), is that during early infancy, human milk or an approved infant formula should be the sole source of nourishment. The panels producing both sets of recommendations emphasized that breastfeeding provides the best milk for the young infant and that if an alternative to this sort of milk has to be introduced, it should be as an approved formula at least up to 6 months of age. The recommendations state, furthermore, that solids may be introduced from about the age of 4 months to complement the milk feeding.

Figures 1 and 2 show secular trends in the incidence of breastfeeding during the 60 year period from 1920 to 1985 as well as changes in the average age for the onset of the introduction of solid foods. Figure 1 reveals a steady decline in breastfeeding for most of this period, both in the scatter of data collated from a series of regional studies in the United Kingdom, as well as from four nationally representative studies conducted by Government agencies. The two notable 'outlyer' data points are for Oxford and Cambridge, emphasizing their suitability for the type of investigation described in this review. Between 1975 and 1980 the two Office of Population

Censuses and Surveys Studies, OPCS (Martin, 1978, Martin & Monk, 1982) do demonstrate, however, a partial reversal of the situation. Although preliminary results from a further OPCS investigation carried out in 1985 indicated that the 1980 position had been maintained, they provided no sign of further improvement.

There have similarly been major alterations in the types of non-human milk fed to babies. Initially, diluted cow's milk to which variable amounts of sugar were added predominated. This practice slowly changed over the next 50 years and by 1970 formula milks were very much the rule. Post-1970 has seen three additional major changes. The formulas are now complete in terms of dietary energy content and no extra sucrose should be added. Greater care has been taken to ensure the feeds are not made up from the powder in a form that contains a higher energy content than human milk, both by improving scoop design and by providing more exact mixing instructions. Finally, either all or most of the animal fat has now been replaced by higher polyunsaturated fat containing vegetable oils. There have been other compositional changes but they are not relevant in the context of this review.

The pattern of change with time in the age when solids were introduced is described in Figure 2. Inspite of continuous advice about the undesirability of giving solids before 4–6 months, the period up to 1975 showed a steady increase in the

Figure 1. Proportions of mothers breastfeeding at 3 months from 1920 to 1985. Closed symbols refer to national surveys • Ministry of Health, 1944, ♦ Douglas, 1950; Martin, 1978; Martin & Monk, 1982. Open symbols refer to individual cities and towns from 29 different studies.

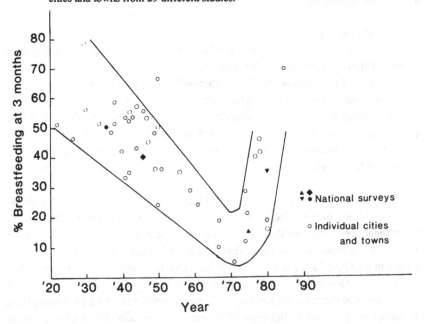

proportion of babies receiving solids at 3 months; indeed DHSS (1974) reported that in the early seventies the great majority had received solids by one month and many as early as two weeks. Firm advice to counter such practices did seem to have some effect between 1975 and 1980 (Martin & Monk, 1982) but again preliminary results from the OPCS (1985) survey indicate only a maintenance of this trend, not its enhancement.

The important connection between the timing of these dietary changes and discussions about growth is that the commonly used British reference standards as well as the American NCHS ones (Hamill, 1977) adopted by WHO were all developed from data collected during periods when dietary practices were substantially different from those we would now recommend. Thus the relevance of these growth reference patterns for assessing growth of infants with currently recommended dietary practices is questionable. The primary concern is not so much for the evaluation of a measurement at a particular age (i.e. distance standards) but for the interpretation of *change* in measurements with time (i.e. growth velocity). Growth curves are universally used by nutritionists and health workers to assess dietary adequacy as well as many other aspects of development: a slippage across the distance reference centiles (representing assessment of velocity) is invariably interpreted as being undesirable.

Figure 2. Age of introduction of solid foods from 1920 to 1980. Solid symbols are means or ranges of reported age of introduction. Open symbols are the mean or range of recommendations given by various authors advising on infant feeding.

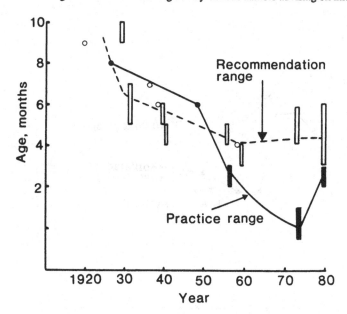

Dietary change and energy intake

Exclusively breastfeeding to 4 months or using carefully constituted milk formulae, plus refraining from solids until after 4 months might all be expected to affect growth, especially the deposition of subcutaneous fat. Figure 3 shows the measured metabolizable energy intakes of Cambridge babies, initially breastfed or given modern formula, in contrast with WHO/FAO (1973) recommended amounts. The latter were based on pre-1970 measured intakes of energy and it is quite clear that the Cambridge values collected in 1980 (Whitehead & Paul, 1981) are some 15–20% lower. Similar values to ours have been observed by many other workers (for review see Whitehead, Paul & Cole, 1981) and both the DHSS (1979) Recommended Dietary Amounts and the FAO/WHO/UNU (1985) Protein and Energy Requirements tables have lowered their energy recommendations to accommodate such findings.

The current Cambridge prospective infant growth study

In order to recruit the necessary babies for a longitudinal study, mothers were approached antenatally through the Cambridge City District Midwives. Babies were recruited in two cohorts, the first 63 between January and March 1984 and the remaining 69 from February to April 1985. Further cohorts are currently being studied but the data are not yet complete. All except five babies of those discussed

Fig. 3. Measured metabolizable energy intakes (kcal/kg body weight) of Cambridge babies, o initially breastfed (n = 48), • bottlefed (n = 9), compared to the WHO/FAO (1973) estimated requirements (dotted lines).

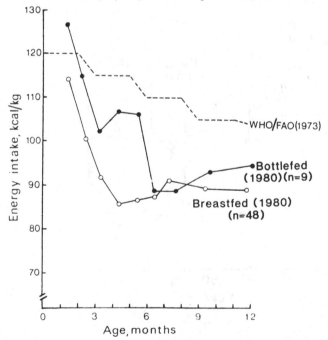

here had both parents of Caucasian origin. Fifty-four percent of the fathers of the baby boys and 62% of those of the girls belonged to the white collar social classes (I, II and III NM) as judged by the Registrar General's classification system.

Anthropometric measurements were carried out (nearly all by the same scientist) every 4 weeks up to 1 year and then at 15, 18, 24, 30 and 36 months. They were made during visits to the infants' own homes in order to minimize drop-outs. The techniques used were in accord with standard approved procedures. Smoothed centiles were calculated from the data by the recently developed method of Cole (1987). At each monthly visit information was also collected on the type of milk being fed as well as on any weaning foods that were also being provided.

Almost 90% of the babies were *initially* breastfed. Only 23% of the boys and 24% of the girls received any formula milk before three weeks of age. Few infants received solids before 10 weeks and the mean age of starting solids was 13.8 ± 4.1 (SD) weeks in the case of boys and 14.4 ± 4.2 weeks with the girls.

Anthropometric results

In situations in which there has been a trend towards a lowered dietary energy intake, one might expect fat stores to be most affected, particularly the subcutaneous fat deposits. Figure 4 shows Cambridge 3rd to 97th centile values for triceps skinfolds and Figure 5 subscapular values. For comparison the Tanner Revised Standards (Tanner & Whitehouse, 1975) are also provided. The triceps exhibit marked differences such that for most of infancy the 50th centile of the Cambridge triceps skinfold lies below the Tanner 10th centile. Subscapular values, although not so dramatically different, still have the Cambridge 50th centile lying between the current 10th and 25th centile values, both for boys and girls.

With such differences in skinfolds it would not be surprising if patterns of weight growth too exhibited marked differences (Figure 6). Interestingly, the Cambridge cohort showed a faster *initial* velocity of growth than either Tanner or NCHS (i.e. Fels Research Institute) reference curves. Whether or not this represents a true difference between 'old' and 'new' infant feeding practices, or merely arises from our curves being fitted through points taken at 4 week intervals and the others at only 12 week intervals, remains to be seen. *After* 3–4 months the Cambridge cohort showed a steady relative decline in weight such that by 10 months the 50th centile Cambridge boy lay nearer the Tanner 25th centile even though at 3 months he had been on the 60th reference centile.

Length exhibited similar trends to weight (Figure 7) but understandably the differences were less extreme than with weight and especially skinfolds.

The fact that growth in weight, length and skinfolds appears to be different now that infant feeding practices have changed is not just a phenomenon seen in Cambridge children; it can also be observed in the published work of a number of other investigators including Butte, Garza, O'Brian-Smith and Nichols (1984) from

Figure 4. 3rd to 97th centiles of triceps skinfold thickness of Cambridge babies (72 boys and 60 girls), solid lines, compared to Tanner centiles (Tanner & Whitehouse, 1975), dotted lines.

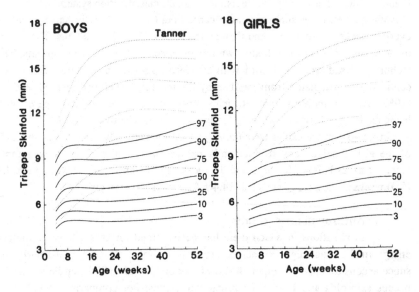

Figure 5. 3rd to 97th centiles of subscapular skinfold thickness of Cambridge babies (72 boys and 60 girls), solid lines, compared to Tanner centiles (Tanner & Whitehouse, 1975), dotted lines.

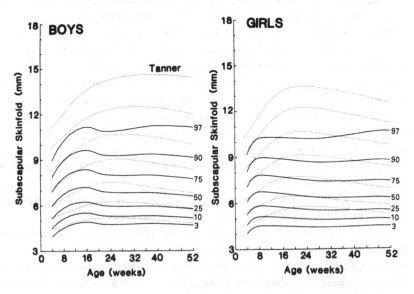

Texas: Owen, Garry & Hooper (1984) in New Mexico: Schluter, Funfack, Pachaly & Weber (1976) from Germany: Boulton (1981) from Australia and Yeung (1983) working in Canada.

Mid upper arm circumference, although a popular and useful measurement in Third World situations, has not previously been studied with the same statistical precision as the other anthropometric criteria. For completeness the Cambridge data have been analysed for this dimension too and the results are given in Figure 8 together with the widely used '100%' line commonly adopted for comparative purposes in the Third World(Jelliffe & Jelliffe, 1969). Once again a fall-off after 4 months is apparent with the Cambridge 50th centile in comparison with the customary reference value.

The only measurement which is dramatically different from the above trends is head circumference (Figure 9). This showed almost no crossing of the centiles over the year and all the centiles were *higher* than those of the corresponding Tanner (1984) reference values. The 50th centile of the Cambridge infants lay approximately on the reference 75th centile (Paul, Ahmed & Whitehead, 1985). Our findings are almost identical with those of Ounsted, Moar & Scott (1985) and are important in that they provide some assurance that the variations in other growth parameters we have observed are not paralleled by any impairment in brain growth and development.

Conclusions

It cannot be claimed with complete certainty that the altered growth patterns described here really do result from dietary changes but such would appear to be the

Figure 6. 3rd to 97th centiles of weight of Cambridge babies (72 boys and 60 girls), solid lines, compared to Tanner centiles (Tanner & Whitehouse, 1973), dotted lines.

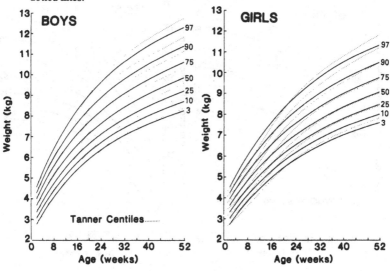

Figure 7. 3rd to 97th centiles of recumbent length of Cambridge babies (72 boys and 60 girls), solid lines, compared to Tanner centiles (Tanner & Whitehouse, 1973), dotted lines.

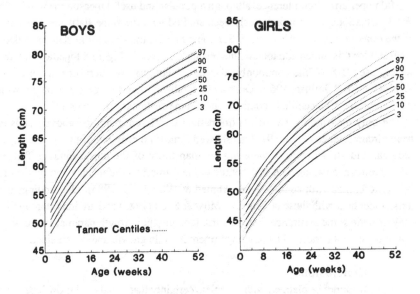

Figure 8. 3rd to 97th centiles of mid-upper arm circumference of Cambridge babies (72 boys and 60 girls), solid lines, compared to the Wolanski Standard (1966) smoothed by the method of El Lozy (1969), dotted line.

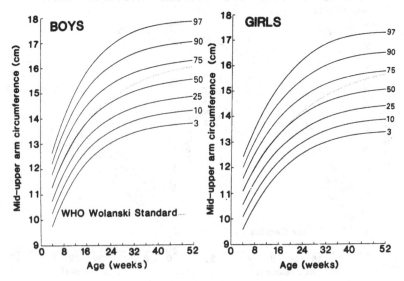

most probable explanation. Be this as it may, it is clearly undesirable to use as growth references, the progress of children who were fed in a manner we would no longer consider desirable. The UK Tanner reference charts for skinfold thickness were, for example, based on measurements carried out in Derbyshire between 1966 and 1967 (Hutchinson-Smith, 1973). Likewise our national weight and height reference curves date from information collected during similar or earlier periods. Growth standards in current use in the United Kingdom include the Gairdner–Pearson charts (Gairdner & Pearson, 1971). In spite of the relatively recent date on the *publication* these authors used the classical Harvard data for head circumference which was collected between 1929 and 1939. The more recently published NCHS references (Hamill, 1977) and now adopted by WHO, contained equally old data including some originating from 1929.

On a number of occasions we have made the plea that it is time that new charts were prepared for the present generation of babies, to describe more adequately the events that are taking place now during the crucial period of growth covered by infancy. The likelihood that such a venture would be scientifically informative is emphasized by the recent Dutch weight and length standards (Roede & Van Wieringen, 1985) which also reflect the smaller size of present day (1980) infants in the Netherlands relative to that which had been described previously. Recognition that the growth status of babies depends on how they are fed will enable a more precise recognition of genuinely abnormal growth and also alleviate anxiety in mothers who are observing 'downward centile crossing' in apparently healthy babies. Though it

Figure 9. 3rd to 97th centiles of head circumference of Cambridge babies (72 boys and 60 girls), solid lines, compared to Tanner centiles (Tanner, 1984), dotted lines.

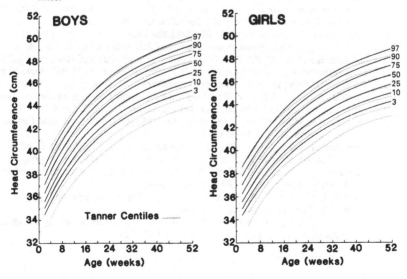

78   R.G. WHITEHEAD *ET AL.*

may be impossible to define the ideal growth target for infants, at least we can make
certain that the guidelines we provide at governmental level for *growth* are compatible
with the advice we give about *diet*.

References

Boulton, J. (1981). Nutrition in childhood and its relationships to early somatic growth, body fat, blood pressure and physical fitness. *Acta Paediatrica Scandinavica*, Supplement 284.

Butte, N.F., Garza, C., O'Brien-Smith, E. & Nichols, B.L. (1984). Human milk intake and growth in exclusively breast-fed infants. *Journal of Paediatrics*, **104**, 187–95.

Cole, T.J. (1987). Fitting smooth centile curves to reference data. *Journal of the Royal Statistical Society A* (in press).

Committee on Medical Aspects of Food Policy, Panel on Child Nutrition (1987). Policy statement on infant feeding. Published in full in *Health Visitor*, **60**, 130.

Department of Health and Social Security (1974). Present-day practice in Infant Feeding. *Report on Health and Social Subjects, Number 9*. HMSO, London .

Department of Health and Social Security (1979). Recommended daily amounts of food energy and nutrients for groups of people in the United Kingdom. *Report on Health and Social Subjects, Number 15*. HMSO, London .

Department of Health and Social Security (1980). Present day practice in infant feeding. *Report on Health and Social Subjects, Number 20*. HMSO, London.

Douglas, J.W.B. (1950). The extent of breastfeeding in Great Britain in 1946, with special reference to the health and survival of children. *Journal of Obstetrics and Gynaecology*, **57**, 335–61.

El Lozy, M. (1969). A modification of Wolanski's standards for the arm circumference. *Journal of Tropical Pediatrics*, **15**, 193–4.

FAO/WHO/UNU (1985). Energy and protein requirements. *World Health Organisation Technical Report Series Number 724*, Geneva, WHO.

Gairdner, D. & Pearson, J. (1971). A growth chart for premature and other infants. *Archives of Disease in Childhood*, **46**, 783–7.

Hamill, P.V.V. (1977). NCHS Growth Curves for Children, birth to 18 years. *US Department of Health, Education and Welfare publication Number PHS 78–1650*, National Centre for Health Statistics, Hyattsville, MD.

Hutchinson-Smith, B. (1973). Skinfold thickness in infancy in relation to birthweight. *Developmental Medicine and Child Neurology*, **15**, 628–34.

Jelliffe, E.F.P. & Jelliffe, D.B. (eds) (1969) The arm circumference as a public health index of protein-calorie malnutrition of early childhood. *Journal of Tropical Paediatrics*, **15**, Monograph number 9, 177–260.

Martin, J. (1978). *Infant feeding 1975: attitudes and practices in England and Wales*. Office of Population Censuses and Surveys. HMSO, London.

Martin, J. & Monk, J. (1982). *Infant Feeding 1980*. Office of Population Censuses and Surveys. HMSO, London.

Ministry of Health (1944). Report on The Breast Feeding of Infants. *Reports on Publications, Health and Medical Subjects, Number 91*. HMSO, London.

Ounsted, M., Moar, V.A. & Scott, A. (1985). Head circumference charts updated. *Archives of Disease in Childhood*, **60**, 936–9.

Owen, G.M., Garry, P.J. & Hooper, E.M. (1984). Feeding and growth of infants. *Nutrition Research*, **4**, 727–31.

Paul, A.A., Ahmed, E.A. & Whitehead, R.G. (1986). Head circumference charts updated. *Archives of Disease in Childhood*, **61**, 927–8.

Roede, M.J. & van Wieringen, J.C. (1985). Growth diagrams 1980. Netherlands third nationwide survey. *Tijdschrift voor Sociale Gezondheidszorg*, **63**, Supplement, 1–34.

Schluter, K., Funfack, W., Pachaly, J. & Weber, B. (1976). Development of subcutaneous fat in infancy. Standards for tricipital, subscapular and suprailiacal skinfolds in German infants. *European Journal of Paediatrics*, **123**, 255–67.

Tanner, J.M. (1984). Physical growth and development. In *Textbook of Paediatrics*, Third edition, Vol I, ed. J.O. Forfar & G.C. Arneil, pp. 278–330. Churchill Livingstone, Edinburgh.

Tanner, J.M., Whitehouse, R.H.& Takaishi, M. (1966). Standards from birth to maturity for height, weight, height velocity and weight velocity: British children 1965. *Archives of Disease in Childhood*, **41**, 613–35.

Tanner, J.M. & Whitehouse, R.H. (1973). Height and weight chart from birth to 5 years allowing for length of gestation. *Archives of Disease in Childhood*, **48**, 786–9.

Tanner, J.M. & Whitehouse, R.H. (1975). Revised standards for triceps and subscapular skinfolds in British children. *Archives of Disease in Childhood*, **50**, 142–5.

Whitehead, R.G. & Paul, A.A. (1981). Infant growth and human milk requirements: a fresh approach. *Lancet*, ii, 161–3.

Whitehead, R.G., Paul, A.A. & Cole, T.J. (1981). A critical analysis of measured food energy intakes during infancy and early childhood in comparison with current international recommendations. *Journal of Human Nutrition*, **35**, 339–48.

WHO/FAO (1973). Energy and protein requirements. *World Health Organisation Technical Report Series Number 522, Food and Agriculture Organisation Nutrition Meetings Report Series Number 52*, Geneva, WHO and Rome, FAO.

Wolanski, N. (1966). Standards for arm circumference, quoted in Jelliffe, D.B. (1966): the assessment of the nutritional status of the community. *World Health Organisation Monograph Number 53*, WHO, Geneva.

Yeung, D.L. (1983). *Infant nutrition: a study of feeding practices and growth from birth to 18 months*. (Ontario: Canadian Public Health Association).

P.J.J. SAUER

Energy requirements and substrate utilization in the newborn infant

The energy requirements for growth can well be studied in the preterm neonate. The rate of weight gain during the last trimester of normal pregnancy is higher than during any other period of human life, 15–17 g/kg/d. The weight of a normal fetus triples in the 10-week period from 30 to 40 wk of normal pregnancy.

At birth a preterm infant is classified, according to the combination of birthweight and gestational age, as appropriate, small or large for gestational age. Several studies have shown metabolic differences between these groups. Measurements of both weight and gestational age however should be considered with caution. An increase in total body water causes an increase in weight. Recently we observed an increase in bodyweight of infants whose mothers received indomethacin to stop labour, the higher weight being probably only caused by an increase in water. Growth therefore is not equal to a gain in bodyweight, and classification as small, appropriate or large for gestational age should really be based on measurements of bodycomposition and not only on measurements of weight, length and gestational age.

Changes in bodycomposition during the last trimester of normal pregnancy are impressive, as shown in Figure 1. The increase in protein is almost linear, while fat is deposited mainly toward the end of pregnancy. The percentage of total body water declines during the last trimester.

It is questionable if the changes in body composition of the fetus are representative of the growing preterm infant. Data on the body composition of the fetus have been collected from infants who died around birth. Secondly, the curves are constructed from cross-sectional data and not from longitudinal studies, which might better represent growth. Finally, the extra-uterine environment is quite different from the intra-uterine environment: the infant is in air as opposed to water, it has to defend itself against changes in environmental temperatures, has to take care of gas-exchange, etc. Energy requirements of growth calculated from changes in bodycomposition of the fetus should therefore be regarded with caution.

The total energy requirement of the fetus, including that for growth, has been estimated in several ways (Sparks, Girard et al., 1980). It is impossible, with present techniques, to measure the oxygen consumption of the human fetus in utero. Root & Root (1923) estimated the oxygen consumption of the human fetus from the increase in oxygen consumption of the mother from the middle of the second trimester to the end of pregnancy and found a value of 5.2 ml/kg/d, equivalent to approximately 146 kJ/kg/d. The oxygen consumption in the human fetus is estimated as 5–8 ml/kg/min (Sparks *et al.*, 1980). This is in accordance with the results of animal studies (Battaglia & Meschia, 1978). The fetal oxygen consumption of a wide variety of mammalian species, including primates, varies within a rather narrow range suggesting that the metabolic rate of the mammalian fetus is relatively independent of fetal size.

The energy requirements of the fetus can also be calculated from the glucose uptake of the fetus, as glucose is the main source of energy. The glucose uptake of the human fetus has not been studied. DiGiacomo *et al.* (1987) studied the glucose uptake and glucose oxidation in the fetal lamb. They found a glucose uptake of 4.44 mg/kg/min and a glucose oxidation of 2.65 mg/kg/min, equivalent to 109.5 and 59.8 kJ/kg/day respectively. This is a slightly lower value than that obtained from oxygen consumption. The difference can be explained from the use of other substrates, lactate and aminoacids. The difference between glucose uptake and glucose oxidation reflects the conversion of glucose into fat in the fetus, followed by fat storage.

The energy requirements for growth should be studied longitudinally in the preterm infant itself as estimations from changes in body composition during fetal life and from oxygen consumption and substrate utilization in utero may not apply to the

Figure 1. Partition of body weight in the human fetus during the third trimester. (After Heim, T., *J. Pediatr. Gastr. and Nutrition*, 1983; 2, 516–41).

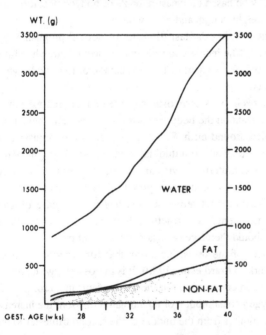

growing preterm infant. Most studies on the energy cost of growth in the preterm infant have used the energy-balance method (Sauer *et al.*, 1985; Brooke *et al.*, 1979; Micheli *et al.*, 1980; Reichman *et al.*, 1981; Rubecz & Mestyan, 1975).

The energy balance can be written as:

energy intake = energy excreted + energy expended + energy components,

where energy intake represents the energy in the food. Energy excreted occurs via faeces and urine. Energy expended includes 1. the energy used for maintenance, 2. the energy used for thermoregulation, 3. the energy used in activity and 4. the energy used for the synthesis of new tissue. Energy components includes the energy present in the components of new tissue. The energy cost of growth is comprised of the energy required for the synthesis of new tissue and the energy stored in the components of this new tissue:

Energy cost of growth = energy components + energy synthesis.

The cost of tissue synthesis represents energy required for the organisation of the components of new tissue and for the formation of complex proteins, lipids, carbohydrates and combinations thereof.

All energy used for maintenance, thermoregulation and activity will finally be given off as heat. None of the energy used for these processes is stored in the body, assuming a stable body temperature. No external work is done by the preterm infant, in contrast to the adult. All energy used by the preterm infant for maintenance, activity and thermoregulation can therefore be measured by calorimetry. The direct measurement of heat transfer (direct calorimetry) is difficult; heat production is therefore mainly calculated by indirect calorimetry. In a steady state situation, all transfer of energy into heat is caused by oxidation, in which oxygen is consumed and carbon dioxide and urea are produced. It is therefore possible to estimate the total heat production from the oxygen consumption, the carbon dioxide production and the urea excretion. This is called indirect calorimetry. As a steady state is needed, indirect calorimetry should always be done over a period of at least 20 minutes.

Very few studies comparing the results of direct and indirect calorimetry have been done (Sauer *et al.*, 1984; Day & Hardy, 1942) in preterm infants. We constructed a new direct and indirect calorimeter (Dane *et al.*, 1985) to measure energy expenditure in infants of less than 2.5 kg. Using this system longitudinal studies were made in 14 preterm infants, birthweight 870–1850 g., gestational age 29–34 wk, age at study 1–58 d, weight 940–2110 g (Sauer *et al.*, 1984). Resting metabolic rate, measured during 20 min of sleep, did not increase with postnatal age after the first week of life (Figure 2). Total metabolic rate (measured by indirect calorimetry) and total heat production (measured by direct calorimetry) both increased with increasing postnatal

Figure 2. Resting metabolic rate as function of postnatal age. (▲) first wk and (●) after first wk. After first wk (y = a ± bx) and (n = 54): a = 245.9 ± 6.4 and b = 0.07 ± 0.2. The slope of the regression line is not statistically different from zero, P < 0.001. (Sauer, Dane & Visser, 1984).

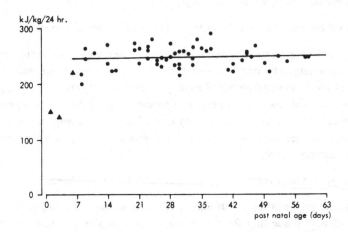

Figure 3.Total metabolic rate and total heat loss as function of postnatal age. (▲) metabolic rate, first wk; (Δ) heat loss, first wk; (●) metabolic rate, áfter first wk; and (O) heat loss, after first wk. After first wk (y = a ± bx) and (n = 54): metabolic rate, a = 269.3 ± 7.3, b = 0.4 ± 0.2; and heat loss, a = 250.0 ± 6.9; b = 0.4 ± 0.2. The slope of both regression lines is statistically different from zero. (Sauer, Dane & Visser, 1984).

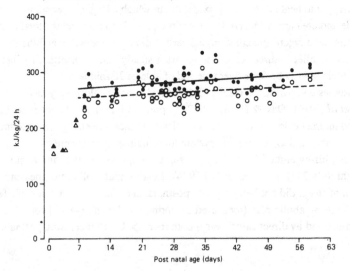

age. The results of indirect calorimetry were higher than those of direct calorimetry in all studies after the first week of life (Figure 3). We hypothesize that this difference is equal to the amount of energy used for tissue synthesis which is stored within the body. Using the data on energy intake and energy losses it is possible to calculate the energy balance (Figure 4). Almost half of the energy taken up by the infant is used for growth. The energy cost per gram weight gain can be calculated from the energy used for growth and weight gain.

A comparison of the results of different studies on the energy requirements for growth is given in Table 1. Studies using the energy balance in preterm infants show some variations in the results. The energy for the components of new tissue is 11.9–16.8 kJ/g weight gain. The energy for synthesis was calculated by Brooke *et al.* (1979) from the difference between the total energy storage and the energy stored in the components of new tissue. In our studies (Sauer *et al.*, 1984) it was calculated from the difference between indirect and direct calorimetry.

Ashworth *et al.* (1968) and Spady *et al.* (1976) calculated the energy requirements for growth during another period of rapid growth, the recovery after malnutrition. Their results are comparable to the results in preterm infants. The energy cost of tissue synthesis was calculated from the increase in heat production after a feed, called the specific dynamic action (SDA). The SDA was high during the period of catch-up growth and decreased after recovery, being correlated to weight gain

Figure 4. Energy balance of a very low birth weight infant, body weight 1–2 kg, growing 17 g.kg^{-1}. 24 h^{-1}. (Sauer, Dane & Visser, 1984).

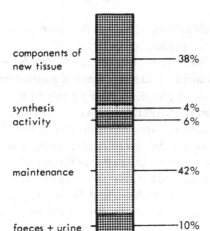

Caloric intake   535 kJ/kg/24 hr

components of new tissue —— 38%

synthesis —— 4%
activity —— 6%

maintenance —— 42%

faeces + urine —— 10%

Table 1. *Energy cost of growth*

Author		Method	kJ/g weight gain		
			E_{comp}	E_{synth}	Total
Gordon *et al.*	1940				9.6
Brooke *et al.*	1979	energy balance	16.8	7.2	24.0
Chessex *et al.*	1981	preterm infants	15.2	1.7	16.9
Sauer *et al.*	1984		11.9	1.1	13.0
Ashworth *et al.*	1968	energy balance			23.0
Spady *et al.*	1976	malnutrition	13.8	4.6	18.4
Widdowson	1974	carcase analysis	6.0	2.5	8.5
Ziegler *et al.*	1976		6.4	3.0	9.4
Brooke *et al.*	1972			1.1	
Ashworth	1972	SDA		1.1	
Chessex	1981			1.7	
Hommes *et al.*	1975	ATP equivalents	6.8	1.3	8.1

(Brooke & Ashworth, 1972). Chessex *et al.* (1981) also used the increase in oxygen consumption after a feed to calculate the energy cost of tissue synthesis.

The energy cost of growth in the energy balance studies is higher than the values which are calculated from the changes in body composition of the fetus in utero. This might be due to differences in the composition of weight gain, as will be discussed later.

The metabolic rate, resting as well as total, is higher in the preterm infant than in children and adults. The metabolic rate in adults is 85–105 kJ/kg/d.

One explanation for the high metabolic rate of the preterm infant might be the rate of protein turnover. Using stable isotope techniques, the rate of protein turnover in preterm infants was found to be 10–16 g/kg/d, depending partly on the route by which the feeding is given (Pencharz *et al.*, 1981; Gatzeflis *et al.*, 1985; Pencharz *et al.*, 1984). Catzeflis *et al.* (1985) calculated from the regression equation of protein synthesis versus energy expenditure that the cost of protein synthesis in preterm infants was 8.35 kJ/g protein. This cost is twice as great as the theoretical cost of protein synthesis calculated from the amount of ATP necessary (4.2 kJ/g protein). However, the cost calculated from stochiometry is minimal. Other energy dissipative mechanisms may be involved during growth and the efficiency of the processes is less than 100%.

Table 2. *Total body protein synthesis and energy metabolism in children and young adults**

Age group	Protein synt g/kg/d	Basal met. rate kcal/kg/d	% of BMR used for protein synthesis**
Prematures	14.4	55	52
Infants	6.1	52.8	23
Children	5.0	40.8	24
Adolescents	5.7	32.4	35
Adults	3.0	24	25

*Ref. Pencharz *et al.*, *Medical Hypothesis*, 1981.
**Assuming an energy cost of protein synthesis of 8.35 kJ/g protein.
(Catseflis *et al.*, *Ped. Res.*, 1985).

Accepting a cost of protein synthesis of 8.35 kJ/g protein, 50% of the metabolic rate in the preterm infant can be explained from protein synthesis. The relationship between metabolic rate and protein synthesis during different stages of life is shown in Table 2. Protein synthesis can account for approximately 25% of the metabolic rate during most parts of life, except for two periods of rapid growth, the preterm period and puberty (Pencharz *et al.*, 1981). Further studies to confirm this finding are needed.

Is growth and body composition influenced by the type and amount of feeding?

Studies on body composition in preterm infants are scarce, as the techniques which are applied in adults cannot be used in the preterm infant. Mattau *et al.* (1977) described a system in which it was possible to measure total body fat from xenon-absorption. No follow-up studies have been published so far. Recently new techniques for measuring body composition have been described: total body conductivity (Cochran *et al.*, 1986), acustic impedance (Deskins *et al.*, 1985) and body volume by air-displacement (Taylor *et al.*, 1985). Longitudinal studies using these techniques have not yet appeared.

Most studies on the energy cost of growth done in the past have used an indirect estimate of the composition of weight gain. It is possible to estimate the composition of weight gain indirectly from the results of indirect calorimetry using the respiratory quotient (RQ)

$$\frac{V_{O_2}}{V_{CO_2}}.$$

When carbohydrate is oxidized, the RQ = 1, compared with RQ = 0.7 when fat is the only substrate used. From the RQ it is possible to calculate the relative contribution of carbohydrate and fat to the heat production. Together with the total oxygen consumption it is possible to calculate the amount of carbyhydrate and fat oxidized. The storage of carbohydrate, fat and protein can be calculated from the difference between intake, losses in urine and faeces and oxidation.

In our study (Sauer *et al.*, 1984) we found that the fat content of new tissue was not correlated with bodyweight and was much higher than the fat accretion in utero (Figure 5). The energy cost of growth was correlated with the fat content of new tissue (Figure 6). A fat content of 0.16 g/g new tissue was found by Mettau using the Xenon absorption technique (Figure 7) (Mettau, 1978). The fat content of weight gain calculated from the intra-uterine accretion between 28 and 32 wks is 0.11 g/g new tissue, compared with 0.20 g/g new tissue in the period of 36–40 wks.

The effect of increasing the energy intake on growth and body composition has been studied (Van Aerde *et al.*, 1985; Reichman *et al.*, 1981). Two groups of preterm infants received either the presently recommended energy intake (500 kJ/kg/d) or a high energy intake (619 kJ/g/d). The protein intake was proportionally increased. Metabolic rate and energy accretion were higher in the group receiving the high energy intake. There was no difference in increase in length or head circumference; weight gain however was higher in the high energy group. The fat content of new tissue was 30% in the high energy group versus 24% in the low energy group, the protein content being not different. This suggests that an increase in energy intake above approximately 500 kJ/kg/d will result in an increase in fat content of new tissue but not in real growth.

Figure 5. Fat content of new tissue as function of body weight. Patients as in Figs 2, 3, 4.

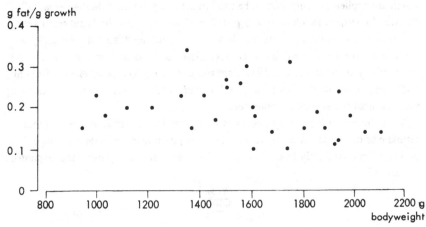

These results are supported by the results of Kashyap *et al.* (1986). In their very elegant study they compared three different but comparable groups of infants. The feeding was different in respect of protein and energy intake (Tables 3, 4). Increased gain of length, head circumference and weight could be achieved by increasing the protein intake from 2.24 g/kg/d to 3.62 g/kg/d at a constant energy intake of 481 kJ/kg/d. Keeping the protein intake at 3.6 g/kg/d and increasing the energy intake to 623 kJ/kg/d resulted in an increase in weight gain but not in gain of length or head circumference. The skinfold thickness increased at the higher energy intake, again suggesting that increased fat deposition occurred at the high energy intake.

All studies mentioned indicate that growth and the composition of new tissue can be manipulated by changing the energy and protein intake of the preterm infant. The composition of the feeding given to the preterm infant might also influence the metabolic rate, weight gain and body composition. The increase of i.v.-administered glucose to term infants caused an increase in energy expenditure, 30% of the increased energy given by carbohydrate being dissipated as heat (Figure 8) (Sauer *et al.*, 1984). This increase in metabolic rate might be explained by an increased conversion of glucose into fat. Replacing part of the glucose calories by fat caused a significant reduction in metabolic rate, leaving more energy for growth (Sauer *et al.*, 1986). No effect on protein turnover was found (Pencharz *et al.*, 1986).

Figure 6. Fat content of new tissue plotted against energy of components. Patients as in Figs 2, 3, 4.

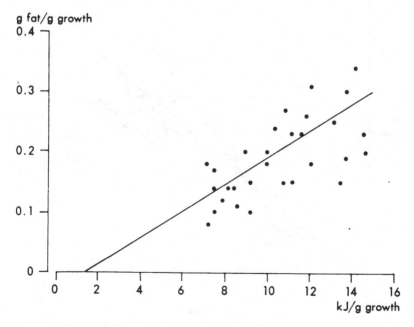

90   P.J.J. SAUER

The result of studies on the effect of differences in the composition of the oral feed on growth and body composition of the preterm infant are somewhat contradictory. Studies in adults suggest that replacing long chain fatty acids by medium chain fatty acids increases the metabolic rate (Geliebter *et al.*, 1983). This could not be confirmed in preterm infants (Whyte & Campbell, 1986). No difference in metabolic rate and weight gain was observed between groups of infants who received either almost all their fatty acids as long chain or a feed in which 40% of the fatty acids were replaced by medium chain fatty acids.

Another comparison was made between two groups of infants of comparable birthweight and gestational age, fed either their mother's own milk or a humanized

Figure 7. Total body fat (TBF) as a function of body weight.

Total body fat (TBF) in grams plotted against body weight (BW) in grams.

Results of 26 measurements in 15 preterm babies (O) and 18 corresponding numbers given to patients in table XI to XVI.

Regression lines are drawn for preterm babies (————) and S.F.D. babies (------) separately. Data derived from the literature (table XXIII) are also given (■), but not included in the calculations for the regression lines.

(Mettau, 1978).

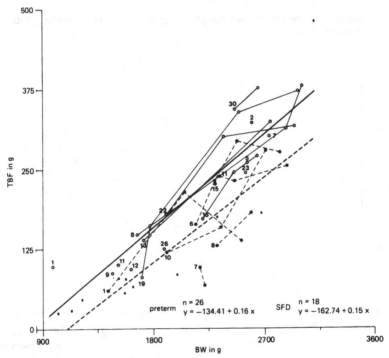

Table 3. *Daily nutrient intake (Kashyap et al., 1986)*

	Group 1 (n = 9)			Group 2 (n = 9)			Group 3 (n = 9)		
Energy (kcal/kg)	115.0	±	0.90	114.0	±	1.00	149.0	±	1.80
Protein (gm/kg)	2.24	±	0.02	3.62	±	0.03	3.5	±	0.04
Fat (gm/kg)	5.9	±	0.05	5.7	±	0.05	7.8	±	0.11
Carbohydrate (gm/kg)	11.6	±	0.12	9.9	±	0.08	14.2	±	0.17
Sodium (mEq/kg)	2.7	±	0.03	2.6	±	0.02	2.6	±	0.03
Potassium (mEq/kg)	4.2	±	0.04	4.2	±	0.03	4.0	±	0.03
Calcium (mEq/kg)	9.9	±	0.10	10.1	±	0.10	10.0	±	0.20
Magnesium (mEq/kg)	1.7	±	0.02	1.5	±	0.01	1.5	±	0.02
Chloride (mEq/kg)	3.6	±	0.04	3.5	±	0.03	3.6	±	0.04
Phosphorus (mg/kg)	99.0	±	1.00	106.0	±	1.00	104.0	±	2.00
Volume (ml/kg)	178.0	±	1.38	178.0	±	1.38	177.0	±	2.26

Values represent mean ± SD.
Source: Kashyap *et al., J. Pediatr.* 1986; **108**, 955–62.

Table 4. *Growth rates (Kashyap et al., 1986)*

	Group 1 (n = 9)	Group 2 (n = 9)	Group 3 (n = 9)
Weight (gm/kg/day)	13.9 ± 2.8[*]	18.3 ± 2.8	22.0 ± 3.1
Length (cm/wk)	0.94 ± 0.19	1.21 ± 0.32	1.24 ± 0.30
Head circumference (cm/wk)	0.85 ± 0.15[*]	1.22 ± 0.28	1.17 ± 0.24
Skinfold thickness			
Triceps (mm/wk)	0.32 ± 0.14	0.38 ± 0.11	0.67 ± 0.34[*]
Subscapular (mm/wk)	0.33 ± 0.20	0.36 ± 0.13	0.64 ± 0.26[*]

Values represent mean ± SD.
[*]Significantly different from other 2 groups.
Source: Kashyap *et al., J. Pediatr.* 1986, **108**, 955–62.

mile (Van Aerde *et al.*, 1987). The energy and protein intake were comparable. The results showed a comparable metabolic rate but a higher fat content of new tissue in the formula fed infants compared with the human milk fed infants.

The fat content of new tissue is higher in all studies of orally fed newborn infants compared with the data on intra-uterine accretion. The percentage of fat in new tissue can be further increased by increasing the energy intake. One explanation for the high fat content of new tissue might be adaptation to extra-uterine life. The fat content of the tissue accretion of preterm infants is comparable with the percentage fat in new tissue found at the end of the normal pregnancy. Heimler *et al.* (1981) showed that the subcutaneous skinfolds increased when preterm infants were transferred from an incubator to a crib. More studies in which the body composition is monitored are needed, but the results obtained so far indicate that the higher fat accretion of the preterm infant compared with the fetus might be an adaptation to extra-uterine life.

Summary and conclusions
One can conclude that the energy requirements for growth can well be studied in the preterm infant. The energy requirements for growth can be divided into energy for the components of new tissue and energy for synthesis. The energy for components is 11.9–16.9 kJ/g growth, for synthesis 1.1–7.2 kJ/g growth.

The composition of weight gain can be influenced by changing the energy intake and the composition of the feed. Increasing the energy intake at a constant protein intake of 3.6 g/kg/d causes a higher fat content of new tissue, but no change in real growth. Increasing the energy intake from 119 to 154 kJ/kg/d with a proportional increase in protein intake from 2.5 to 3.3 g/kg/d resulted also in an increased fat

Figure 8. Glucose oxidation vs. glucose intake.

(Sauer *et al.*, 1984).

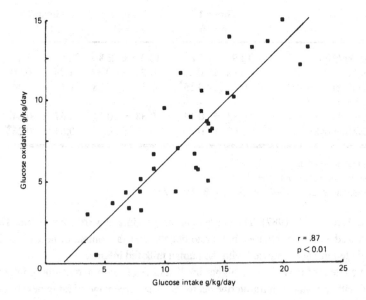

content of new tissue without improving the gain in length and head circumference. Further studies, measuring directly the changes in body composition, are needed.

References

Ashworth, A., Bell, R., James, W.P.T. & Waterlow, J.C. (1968). Calorie requirements of children recovering from protein–calorie malnutrition. *Lancet*, i, 600–3.

Battaglia, F.C. & Meschia, G. (1978). Principal substrates of fetal metabolism. *Physiol. Rev.*, **58**, 499–527.

Brooke, O.G., Alvear, J. & Arnold, M. (1979). Energy retention, energy expenditure, and growth in healthy immature infants. *Ped. Res.*, **13**, 215–20.

Brooke, O.G. & Ashworth, A. (1972). The influence of malnutrition on the postprandial metabolic rate and respiratory quotient. *Br. J. Nutr.*, **27**, 407–15.

Catzeflis, C., Schutz, Y., Micheli, J.L., Welsch, C., Arnaud, M.J. & Jequier, E. (1985). Whole body protein synthesis and energy expenditure in very low birth weight infants. *Ped. Res.*, **19**, 679–87.

Chessex, P., Reichman, B.L., Verellen, G.J.E., Putet, G., Smith, J.M., Heim, T.& Swyer, P.R. (1981). Influence of postnatal age, energy intake, and weight gain on energy metabolism in the very low-birth-weight infant. *J. Pediatr.*, **99**, 761–6.

Cochran, W.J., Klish, W.J., Wong, W.W. & Klein, P.D. (1986). Total body electrical conductivity used to determine body composition in infants. *Ped. Res.*, **20**, 561–4.

Dane, H.J., Holland, W.P.J., Sauer, P.J.J. & Visser, H.K.A. (1985). A calorimetric system for metabolic studies of newborn infants. *Clin. Phys. Physiol. Meas.*, **6**, 37–46.

Day, R. & Hardy, J.D. (1942). Respiratory metabolism in infancy and in childhood XXVI A calorimeter for measuring the heat loss of premature infants. *Am. J. Dis. Child*, **63**, 1086–95.

Deskins, W.G., Winter, D.C., Sheng, H.P. & Garza, C. (1985). An acoustic plethysmograph to measure total infant body volume. *J. Biomech. Eng.*, **107**, 304–8.

DiGiacomo, J.E., Hay, W.W. Jr & Battaglia, F.C. (1987). Effect of increasing glucose concentration alone on fetal glucose utilization. *Ped. Res.*, **21**, 340A.

Geliebter, A., Torbay, N., Bracco, E.F., Hashim, S.A.& Van Itallie, T.B. (1983). Overfeeding with medium-chain triglyceride diet results in diminished deposition of fat. *Am. J. Clin. Nutr.*, **37**, 1–4.

Heimler, R., Sumners, J.E. Grausz, J.P., Kiern, C.L. & Glaspey, J.C. (1981). Thermal environment to change in growing premature infants: effect on general somatic growth and subcutaneous fat accumulation. *Pediatrics*, **68**, 82–86.

Koshyap, S., Forsyth, M., Zucker, C., Ramakrishnan, R., Dell, R.B. & Heird, W.C. (1986). Effects of varying protein and energy intakes on growth and metabolic response in low birth weight infants. *J. Pediatr.*, **108**, 955–63.

Mettau, J.W. (1978). Measurement of total body fat in low birthweight infants. Thesis, Rotterdam.

Mettau, J.W., Degenhart, H.J., Visser, H.K.A. & Holland, W.P.J. (1977). Measurement of total body fat in newborns and infants by absorption and desorption of non-radioactive xenon. *Ped. Res.*, **11**, 1097.

Micheli, J., Gudinchet, F., Stettler, E., Schutz, Y. & Jequier, E. (1980). Relationship between energy expenditure and weight gain in very low birth weight infants during the first week of life. *Ped. Res.*, **14**, 1421.

Pencharz, P.B., Beesley, J., Canagarayar, U., Van Aerde, J., Renner, J., Souer, P., Wesson, D. & Swyer, P. (1984). Protein metabolism of parenterally fed neonates: combined 13C and 15N studies. *Ped. Res.*, **18**, 208A.

Pencharz, P.B., Beesley, J., Sauer, P.J.J., Van Aerde, J., Canagarayer, U., Renner, J., McVey, M., Wesson, D. & Swyer, P. (1986). Effects of energy sources on protein synthesis and substrate utilization of parenterally fed neonates. *Ped. Res.*, **20**, 416A.

Pencharz, P., Masson, M., Desgranges, F. & Papageorgiou, A. (1981). Total-body protein turnover in human premature neonates: effects of birth weight, intra-uterine nutritional status and diet. *Cl. Sci.*, **61**, 207–15.

Pencharz, P.B., Parsons, H., Motil, K. & Duffy, B. (1981). Total body protein turnover and growth in children: is it a futile cycle? *Med. Hypoth.*, 7, 155–60.

Reichman, B., Chessex, P., Putet, G., Verellen, G.J.E., Smith, J.M., Heim, T. & Swyer, P. (1981). Diet, fat accretion and growth in premature infants. *N. Engl. J. Med.*, **305**, 1495–1500.

Root, H.F. & Root, H.K. (1923). *Arch. Intern. Med.*, **32**, 411–24.

Rubecz, I. & Mestyan, J. (1975). The partition of maintenance energy expenditure and the pattern of substrate utilization in uterine malnourished newborn infants before and during recovery. *Acta Paediatr. Acad. Sci. Hung*, **16**, 335–50.

Sauer, P.J.J., Dane, H.J. & Visser, H.K.A. (1984). Longitudinal studies on metabolic rate, heat loss, and energy cost of growth in low birth weight infants. *Ped. Res.*, **18**, 254–9.

Sauer, P.J.J., Van Aerde, J., Heim, T. & Swyer, P.R. (1984). Do newborn infants have a limited glucose oxidation rate? *Crit. Care Med.*, **12**, 337.

Sauer, P.J.J., Van Aerde, J., Pencharz, P., Smith, J., Heim, T., Filler, R. & Swyer, P. (1986). Beneficial effect of the lipid system on energy metabolism in the intravenously alimented newborn infant. *Ped. Res.*, **20**, 248A.

Spady, D.W., Payne, P.R., Picou, D. & Waterlow, J.C. (1976). Energy balance during recovery from malnutrition. *Am. J. Clin. Nutr.*, **29**, 1073–8.

Sparks, J.W., Girard, J.R. & Battaglia, F.C. (1980). An estimate of the caloric requirements of the human fetus. *Biol. Neonate*, **38**, 113–9.

Taylor, A., Aksoy, Y., Scopes, J.W., du Mont, G. & Taylor, B.A. (1985). Development of an air displacement method for whole body volume measurement of infants. *J. Biomed. Eng.*, 7, 9–17.

Van Aerde, J., Sauer, P.J.J., Heim, T., Smith, J. & Swyer, P. (1985). Growth, macronutrient oxidation and accretion in very low birthweight (VLBW) infants with variable energy intake and constant diet composition. *Ped. Res.*, **19**, 368A.

Van Aerde, J., Sauer, P., Heim, T., Swyer, P.A. & Smith, J. (1987). Comparison of long-chain-triglyceride (LCT) formula vs. own mother's milk (OMM) feeding on growth, macronutrient and energy balance in very low birth weight (VLBW) infants. *Ped. Res.*, **21**, 439A.

Whyte, R.K., Campbell, D., Stanhope, R., Bayley, H.S. & Sinclair, J.C. (1986). Energy balance in low birth weight infants fed formula of high or low medium-chain triglyceride content. *J. Pediatr.*, **108**, 964–71.

Widdowson, E.M. (1974). Changes in body proportions and composition during growth. In *Scientific Foundation of Paediatrics*, ed. J.A. Davis & J. Dobbing, pp. 153–63. Heineman, London.

Ziegler, E.E., O'Donnell, A., Nelson, S.E. & Fomon, S.J. (1976). Body composition of the reference fetus. *Growth*, **40**, 329–41.

P.S.W. DAVIES & M.A. PREECE

Body composition in children: methods of assessment

Introduction

There are a number of specific areas in the study of child growth and development and paediatric medicine where accurate and reliable knowledge of body composition would be advantageous. In the study of the physical growth and development of children and adolescents anthropometric techniques such as measuring stature and sitting height have long been used to obtain information on linear growth of the child. Detailed knowledge regarding the growth of the tissue compartments of the body, for example, body fat, water, and the lean body mass has been less easily forthcoming using these techniques. Changes in body weight tell us little about the growth of individual tissues. The use of skinfold calipers for the measurement of skinfold thickness augments other anthropometric measurements. However, there is increasing recognition that skinfold measurements and moreover their extrapolation to measures of total body fatness are fraught with problems and assumptions that are not easily overcome or validated (Martin *et al.*, 1985).

To understand and describe the changes that occur in body composition throughout growth would certainly give a new insight into the growth process. There is still much to be learnt of the effect of different hormones on specific body tissues and of the influence of body composition abnormalities in early life on subsequent growth and development. The possible role of body composition in pubertal and reproductive events is still also still far from clear (Frisch & McArthur, 1974; Cameron, 1976; Frisch, 1984; Malina, 1983).

Moreover, there are situations where disturbances in body composition are of clinical importance to the paediatrician. For example, the characteristic changes in body composition experienced by many individuals when receiving corticosteroids as treatment for a variety of inflammatory conditions, i.e. rapid weight gain and changes in fat distribution (Horber *et al.*, 1986) can often be more disturbing to children and adolescents than the disease process itself.

There are also marked body composition changes in some endocrine disorders, such as Cushing's disease and growth hormone deficiency. Whilst the effects of the initiation of growth hormone treatment on body fat and muscle have been investigated (Tanner *et al.*, 1971; Collip *et al.*, 1973; Parra *et al.*, 1979) the potentially more important changes that occur in the quantity and histology of body muscle and fat after the cessation of treatment are only now being addressed (Preece, Round & Jones, 1987).

The study of human body composition is confounded by the obvious necessity of using indirect techniques. Also, there are a very limited number of body composition techniques that allow individual tissues or body constituents to be assessed. The vast majority of available indirect techniques consider the human body as a two compartment model consisting of fat and fat free mass. The fat free mass therefore consists of many different tissues, inter alia bone muscle and the viscera. The number of techniques of body composition assessment appropriate to paediatrics are less than those available when studying the adult human because of special problems encountered in this field. Ethical considerations reduce the number of techniques that can be used. For example, the use of radioactive tritium oxide for the assessment of total body water is usually precluded. Also the detailed analysis of body composition using in vivo neutron activation (Allen, Gaskin & Stewart, 1986) has important ethical considerations because of exposure to radiation although the incurred radiation dose is relatively small, being about 30 cGy (Forbes, 1978). The recent development of photon activation analysis (Ulin et al., 1986) for the measurement of total body carbon, oxygen and nitrogen has yet to be applied to paediatric body composition analysis, and again the radiation dose incurred may be restrictive.

No less an important consideration when studying paediatric body composition is that technique should be acceptable to the child. Thus the procedure of measuring body density via underwater weighing or volumetric displacement, while being non-invasive, can prove difficult to use in children, especially sick children. Also the measurement of the naturally occurring isotope potassium^{40} as an index of the lean body mass can prove disturbing to many children because of the claustrophobic nature of the counting chamber.

The repeatability and reliability of any potential body composition technique should be carefuly evaluated and this is no less important in paediatric studies than in adult body composition analysis.

Finally techniques for the assessment of body composition should ideally be applicable in both health and disease. A disturbance in body fat distribution, or an unusual distribution in some diseases, such as Cushing's disease may for example affect the ability of skinfold caliper measurements to predict total body fatness.

Subcutaneous fat measurement

Thus ethical considerations and some of the problems outlined above all make the achievement of accurate analysis of paediatric body composition difficult. One approach to paediatric body composition that has been used by many workers has been to consider the changes that occur throughout childhood and adolescence in skinfold thickness. This technique satisfies many of the above criteria by being non-invasive and in the most part acceptable to the child. Figure I shows the logged sum of the mean thickness of the two most frequently taken skinfolds, namely at the triceps and subscapular sites, at different ages, in boys and girls. These curves have been taken by many to be representative of changes that occur in total body fat throughout the period of growth (Brook, 1978; Fomon *et al.*, 1982). Indeed the curves show many of the features thought to be characteristic of changes in total body fatness that occur in the normally growing child. An initial rapid rise in both sexes in the first six months or so of life is followed by a gradual decline which is more marked in boys. In both sexes a nadir is reached at about seven years of age, after which there is a gradual increase in values, with a characteristic dip in boys at about the time of the adolescent growth spurt. This temporary reduction in fat at that time is a function of the triceps skinfold only. Although there is a slowing in the rate of increase of subscapular measurements at about the age of thirteen years there is no actual reduction in the 50th centile value (Tanner & Whitehouse, 1976). The methodology of monitoring skinfold thickness changes longitudinally throughout growth has been a popular approach to determining total body fatness and hence lean tissue mass. There are many equations that purport to predict total body fat from skinfold caliper measurements both alone and in conjunction with other easily measured anthropometric variables such as height and weight (Parizkova, 1961; Durnin & Rahaman, 1967; Brook, 1971; Lohman, Boileau & Massey, 1975; Frerichs, Harsha & Berenson, 1979). Indeed, over 100 prediction equations have been produced in the last 40 years. There is however an increasing awareness that the

Figure 1. The logged sum of the mean thickness at the triceps and subscapular skinfold sites at different ages.

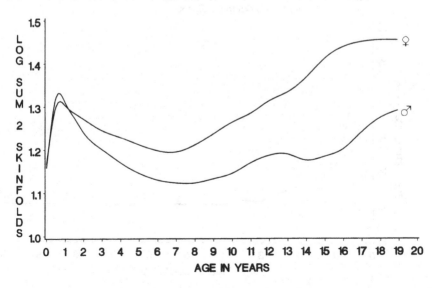

use of skinfold calipers has a number of problems associated with it that are not easily overcome (Martin *et al.*, 1985).

Initially there is the problem of skinfold compressibility. This phenomenon has long been appreciated (Brozek & Kinsey, 1960); Figure 2 shows the compression of a skinfold taking place over a period of ten seconds after the application of the skinfold caliper. The measurements were taken using the Harpenden electronic read-out (HERO) skinfold caliper (Jones, Marshall & Branson, 1979), at the suprailiac site of an adolescent girl. The rapid compression of the skinfold in the first few seconds can be seen clearly. In the neonate such compressive changes continue for as long as sixty seconds after the application of the skinfold caliper (Brans *et al.*, 1974). It is well documented that the amount of compression varies between individuals and at different skinfold sites (Brown & Jones, 1977; Jones, Davies & Norgan, 1986) and it is probable that the rate of compression varies also (Davies, 1987 unpublished results). Consequently, the same actual thickness of subcutaneous adipose tissue at any site in two individuals may well yield different skinfold caliper measurements. Measurement of subcutaneous adipose tissue thickness by A-mode ultrasound overcomes the problems of compressibility and has been shown to improve the accuracy of prediction of total body fat from subcutaneous measures (Sloan, 1967; Jones *et al.*, 1986).

The choice of representative body sites at which to obtain measurements of skinfold thickness is equally problematic. Sites should be used that satisfy the criteria of Lewis, Masterton & Ferres (1958). Their criteria for selection of six sites to represent the entire body were 'practicability, anatomical representativeness,

Figure 2. The compression of a suprailiac skinfold measurement over ten seconds as measured using HERO skinfold calipers.

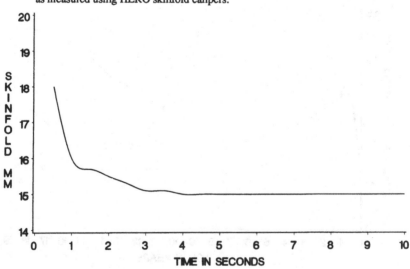

sensitivity to change in fat deposition and reproducability'. The sites most often measured and included in predictive equations are at the triceps, biceps, subscapular and suprailiac sites. It is interesting to note that the majority of equations used to predict total body fatness from subcutaneous measures have no contributory skinfold from the lower limb. It is well accepted that skinfold measurements on the thigh are often the most significant in the prediction of total body fat in adults (Lohman, 1981) and can be one of the major areas of subcutaneous fat deposition especially in females (Satwanti, Singh & Bharadwaj, 1980; Brown & Jones, 1977). Despite this, measurements of thigh subcutaneous fat are rarely included in predictive equations devised specifically for children.

Skinfold thicknesses are notoriously difficult to measure accurately. A high degree of training and constant practice is required if an individual's reliability and repeatability is to be maintained. Subject co-operation assists in the accurate measurement of a skinfold and in the very young this co-operation is often absent. In older children the reliability of skinfold measurements would seem to be partially dependent on age, with higher measurement errors apparent in the older child (Johnston & Mack, 1985). This observation can be accounted for by two facts. Firstly, there is an increase in the variance of skinfold measurements with age associated with increasing values of the mean. Secondly, it is more difficult to obtain accurate measurements in large skinfolds. Thus, in the obese and in the very lean, skinfolds can be difficult to measure accurately. These types of individual are ironically often the very groups of individuals for whom accurate body composition data is required.

One further reason for the inability of skinfold measurements to predict total body fat with a high degree of accuracy is the variability between subcutaneous and internal body fat stores (Davies, Jones & Norgan, 1986). All equations that relate skinfold caliper measurements to total body fat assume that there is a constant relationship between subcutaneous and internal fat at any given level of fatness. There is no data available as yet in which the relationship between internal and external body fat has been studied in children or adolescents. Nevertheless, there is published data from an indirect assessment in a group of young adults (Davies *et al.*, 1986) and a rare direct study involving cadaverous material (Martin *et al.*, 1985). This later study forms a part of the important Brussels cadaver analysis study. Total body fat was carefully dissected from thirteen unembalmed cadavers and fat was divided into subcutaneous and internal stores. For the females (N = 7) the correlation coefficients between internal and subcutaneous fat mass was 0.89 (p < 0.01) while for the males (N = 6) the correlation coefficient was less significant (r = 0.75, p < 0.1). Martin and his co-workers conclude that the slope of the regression lines relating subcutaneous and internal fat masses were similar in the sexes and each kilogram of subcutaneous body fat deposited is associated with the accumulation of about 200 g of internal fat. This concept of a linear relationship between the body's two major fat stores is in marked contrast to the results of the study involving young adults. In this study total body fat was measured using densitometry, while subcutaneous fat mass was determined using the mean thickness of subcutaneous fat over the body's surface in conjunction with body surface area. Internal fat mass was calculated as the difference between total body fat mass and subcutaneous fat mass. These workers reported very poor correlation coefficients between subcutaneous fat mass and internal fat mass in 25 women (r = –0.01) and 21 men (r = 0.05). The wide range of values found in this study is illustrated in Figure 3.

This aspect of body composition in the child deserves and is receiving further attention, not only because of the possible consequences for the prediction of total body fat from subcutaneous measures but also due to the burgeoning interest in the relationship between fat distribution and disease.

All the problems described confound the ability of skinfold caliper measurements to predict total body fat with the degree of accuracy one would wish. Examples of the different dependent and independent variables used in predictive equations and in the coefficient of determination of each equation are shown in Figure 4. Notable are the low R^2 values for three of the four equations and the lack of improvement in prediction accuracy when more skinfolds are used, as in the equation of Durnin & Rahaman (1967). Some equations have used as many as 28 independent variables

Figure 3. The variation in the relationship between internal fat mass and subcutaneous fat mass in a group of healthy young adults. Females shown as squares, males as stars.

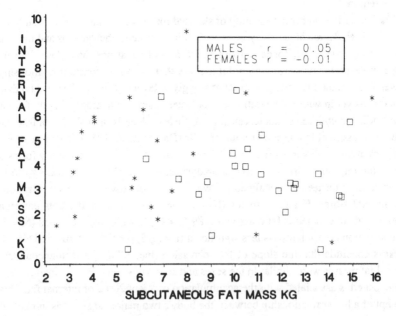

without a marked improvement in the accuracy of the prediction (Young, Sipin & Roe, 1968). Finally, depending on which equation is used the same anthropometric data can produce considerably different answers in terms of percentage body fat, inferring that the equations are highly population specific.

Techniques dependent on body water

In the last few years a number of new body composition techniques have been developed that may have a role to play specifically in paediatric body composition analysis. These include the measurement of total body electrical conductivity (Cochran *et al.*, 1986; van Loan & Mayclin, 1987) and the measurement of total body water using total body electrical impedance (Lukaski *et al.*, 1985; Kushner & Schoeller, 1986). The latter technique, although the subject of a certain amount of criticism in the literature (Cohn, 1985) is gaining popularity worldwide.

The technique of measuring total body electrical impedance is remarkable if only for its simplicity. More importantly, the technique is non-invasive and rapid, the equipment required is relatively cheap and the methodology highly acceptable. These factors therefore make the measurement of total body impedance as an index of total body water an ideal technique for paediatric body composition analysis.

The term total body impedance refers to the electrical resistance offered by the human body when an alternating electrical current is passed through the body. The term impedance usually refers to a 'resistance' due to the combination of simple electrical resistors and capacitors which offer their own resistance in the form of capacitive reactance. Therefore, although there are differences between the exact definitions of the terms resistance and impedance, empirical testing (Lukaski *et al.*, 1985) has shown that within the human the resistance component is the most important. There are a number of factors that influence the resistance of any

Figure 4. Examples of some of the variables used to predict indices of body composition in children and adolescents. The R^2 value of the predictive equation is also shown.

	PREDICTS	USING		R^2
Parizkova, 1961	**Body Density**	**Triceps**	**Subscapular**	0.79
Lohman, 1975	**Lean Body Mass**	**Triceps** **Weight**	**Subscapular**	0.92
Frerichs, 1979	**% Body Fat**	**Triceps** **Height**	**Subscapular**	0.72
Durnin, 1967	**Body Density**	**Triceps** **Biceps**	**Subscapular** **Suprailiac**	0.58

conducting body, including the human, to an alternating electrical current. These are (1) conductor length; (2) conductor configuration; (3) frequency of the current; and (4) conductor cross-sectional area. Using a constant signal frequency and assuming a relatively constant conductor configuration the mathematical equation relating these parameters to resistance is given by:

$$\text{Resistance} = \frac{\text{Length}}{\text{Cross-sectional area}} \qquad (1)$$

If one multiplies by length/length the relationship becomes

$$\text{Resistance} = \frac{\text{Length}^2}{\text{Volume}} \qquad (2)$$

and simple rearrangement gives

$$\text{Volume} = \frac{\text{Length}^2}{\text{Resistance}} \qquad (3)$$

Conductor length is taken to be the height of the individual, and thus the volume of the conducting medium, being total body water, is proportional to a simple relationship involving only height and resistance. This is where the potential of the technique becomes apparent.

Knowledge of total body water is a useful body composition parameter *per se*, however its usefulness to body composition analysis in the broader sense lies in its application in predicting lean body mass and hence fat mass. It is accepted that fat is anhydrous and that the water content of the lean tissue is remarkably constant (Forbes, 1962; Behnke & Wilmore, 1974). Indeed, the concept of a consistant composition of the lean body mass is essential to most indirect body composition techniques. Thus if body water can be measured using the electrical resistance of the body it is possible to derive values for the lean body mass and fat mass.

The anhydrous nature of body fat restricts the flow of current through the human to the lean body mass. The passage of current through the lean tissue has been described as being analogous to the circuit diagram shown in Figure 5. Extracellular water offers a simple resistance to the flow of current (R_e). Equally there is a resistance offered to the current due to intracellular water (R_i). Cell membranes also have a resistance (R_m) and also act as small imperfect capacitors, offering a resistance in the form of capacitive reactance. At high alternating current frequencies the reactive capacitance of the cell membrane becomes very small as it is related to the reciprical of current frequency, effectively short circuiting the membrane resitance. Consequently there is a flow of alternating current through both the extracellular and intracellular water compartments.

The concept that the electrical resistance of the human body is related to its water content is far from new. In the early 1960s Thomasset wrote extensively on the theoretical aspects of the relationship (Thomasset, 1962; Thomasset, 1963). This early work laid the foundations of much of the more recent interest in the technique.

However it was several years after Thomasset's major works that the first empirical testing was carried out. Hoffer and his colleagues in America measured total body water using a standard tritium oxide dilution technique and total body resistance using a simple electrical apparatus in twenty normal adults (Hoffer, Meador & Simpson, 1969). They found a high degree of correlation (r = 0.92, p < 0.001) between total body water in litres and height^2/resistance (H^2/R). Encouraged by these results the procedure was repeated on a group of patients with a variety of pathological conditions including congestive heart disease and chronic renal disease. Once again a high correlation was found between total body water and H^2/R (r = 0.93, p < 0.001). Nevertheless, after this initial and encouraging early work there was a lull of some sixteen years before the technique was investigated further. In 1985 Lukaski and colleagues (Lukaski *et al.*, 1985) measured total body water using tritium oxide dilution in thirty-seven healthy young men. Body impedance was measured using a constant 800 microamp, 50Khz alternating current and recording the voltage drop across the body. Using Ohm's Law resistance was calculated. A highly significant correlation coefficient was found between total body water and H^2/R (r = 0.95, p < 0.0001). More recently Kushner and Schoeller (1986) repeated this work in a group of 58 subjects and once again a highly significant correlation coefficient was found between the two parameters (r = 0.97, p > 0.0001).

Preliminary work relating total body water to impedance in children and adolescents (Davies, Hicks, Halliday & Preece, unpublished results) is encouraging. Total body water has been measured in twenty-six children with a variety of pathological conditions including inflammatory bowel disease, growth hormone deficiency and diabetes, using the stable isotope of water, H_2O^{18}. The use of this

Figure 5. Schematic circuit diagram representing the various electrical elements in the fat free mass.

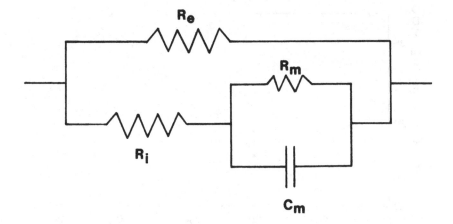

isotope of water is advantageous as it is thought to be a more accurate method of total body water assessment than the use of tritiated or deuterated water as there is less non-aqueous exchange of the labelled atom. Moreover, the use of H_2O^{18} in children is preferable as total body water can be assessed by analysis of expired breath samples, and thus there is no necessity for venepuncture or urine collection (Schoeller *et al.*, 1980). Body resistance was measured by applying a constant 800 microamp 50 Khz alternating current across the left wrist and ankle using non-polarising skin electrodes. Voltage drop was simultaneously measured using two further skin electrodes placed slightly proximal to the current inducing electrodes. Resistance was calculated using Ohm's Law. All the children found the procedure acceptable and of course painless. Time taken to carry out one measurement is about two minutes. The relationship between total body water and height[2]/resistance found in the study group is shown in Figure 6. The high correlation coefficient indicates that over a wide age range H^2/R can accurately predict total body water in children and adolescents. The similarity between the regression equation found in this study and other works (Hoffer *et al.*, 1969; Lukaski *et al.*, 1985; Kushner & Schoeller, 1986) is also encouraging. Further work is being undertaken to examine the ability of impedance to predict total body water in normal children.

In conclusion, to further our knowledge of body composition changes that occur both in normal and abnormal conditions during childhood and adolescence we should fully explore the many new techniques that are being developed. One such is the

Figure 6. The relationship between height[2]/resistance and total body water in a group of 26 children and adolescents.

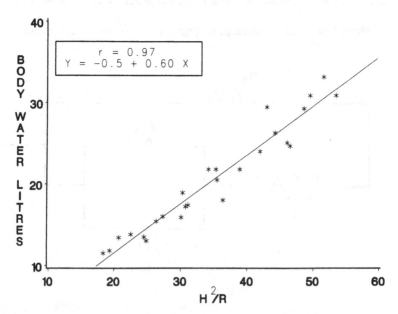

measurement of body impedance. We may then be able to answer more fully and more easily many of the questions concerning body composition pertinent to the human biologist, nutritionist and clinician.

References

Allen, B., Gaskin, K. & Stewart, P. (1986). Measurement of body composition by in vivo neutron activation. *Medical Journal of Australia*, **145**, 307–8.

Behnke, A.R. & Wilmore, J.H. (1974). *Evaluation and Regulation of Body Build and Composition*. Prentice Hall, Inc., New Jersey.

Brans, Y.W., Sumners, J.E., Dweck, H.S. & Cassady, G. (1974). A noninvasive approach to body composition in the neonate. Dynamic skinfold measurements. *Pediatric Research*, **8**, 215–22.

Brook, C.G.D. (1971). Determination of body composition of children. *Archives of Disease in Childhood*, **46**, 182–4.

Brook, C.G.D. (1978). Cellular growth: Adipose tissue. In *Human Growth*, 2nd edn, ed. F. Falkner & J.M. Tanner, Vol. 2. Plenum, New York.

Brown, W.J. & P.R.M. Jones. (1977). The distribution of body fat in relation to habitual activity. *Annals of Human Biology*, **4**, 537–50.

Brozek, J. & Kinzey, W. (1960). Age changes in skinfold compressibility. *Journal of Gerontology*, **15**, 45–51.

Cameron, N. (1976). Weight and skinfold variation at menarche and the critical body weight hypothesis. *Annals of Human Biology*, **3**, 279–82.

Cochran, W.J., Klish, W.J., Wong, W.W. & Klein, P.D. (1986). Total body electrical conductivity used to determine body composition in infants. *Pediatric Research*, **20**, 561–5.

Cohn, S.H. (1985). How valid are bioelectrical impedance measurements in body composition studies? *American Journal of Clinical Nutrition*, **42**, 889–90.

Collipp, P.J., Curti, V., Thomas, J., Sharma, R.K., Maddaiah, V.T. & Cohn, S.H. (1973). Body composition changes in children receiving human growth hormone. *Metabolism*, **22**, 589–98.

Davies, P.S.W., Jones, P.R.M. & Norgan, N.G. (1986). The distribution of subcutaneous and internal fat in man. *Annals of Human Biology*, **13**, 189–92.

Durnin, J.V.G.A. & Rahaman, M.M. (1967). The assessment of the amount of fat in the human body from measurements of skinfold thickness. *British Journal of Nutrition*, **21**, 681–9.

Fomon, S.J., Haschke, F., Ziegler, E.E. & Nelson, S.E. (1982). Body composition of reference children from birth to age 10 years. *American Journal of Clinical Nutrition*, **35**, 1169–75.

Forbes, G.B. (1962). Methods for determining composition of the human body. *Pediatrics*, **29**, 477–94.

Forbes, G.B. (1978). Body composition in adolescence. In *Human Growth*, 2nd edn, ed. F. Falkner & J.M. Tanner, Vol. 2. New York, Plenum.

Frerichs, R.R., Harsha, D.W. & Berenson, G.S. (1979). Equations for estimating percentage body fat in children 10 to 14 years old. *Pediatric Research*, **13**, 170–4.

Frisch, R.E. (1984). Body fat, puberty and fertility. *Biological Reviews*, **59**, 161–88.

Frisch, R.E. & McArthur, J. (1974). Menstrual cycles: fatness as a determinant of minimum weight for height necessary for their maintenance or onset. *Science*, **185**, 949–51.

Hoffer, E.C., Meador, C.K. & Simpson, D.C. (1969). Correlation of whole-body impedance with total body water volume. *Journal of Applied Physiology*, **27**, 531–4.

Horber, F.F., Zurcher, R.M., Herren, H., Crivelli, M.A., Robolti, G. & Frey, F.J. (1986). Altered body fat distribution in patients with glucocorticoid treatment and in patients on long term dialysis. *American Journal of Clinical Nutrition*, **43**, 758–70.

Johnston, F.E. & Mach, R.W. (1985). Interobserver reliability of skinfold measurements in infants and young children. *American Journal of Physical Anthropology*, **67**, 285–9.

Jones, P.R.M., Davies, P.S.W. & Norgan, N.G. (1986). Ultrasonic measurements of subcutaneous adipose tissue thickness in man. *American Journal of Physical Anthropology*, **71**, 359–65.

Jones, P.R.M., Marshall, W.A. & Branson, S.J. (1979). Harpenden electronic read-out (HERO) skinfold calipers. *Annals of Human Biology*, **6**, 159–62.

Kushner, R.F. & Schoeller, D.A. (1986). Estimation of total body water by electrical impedance analysis. *American Journal of Clinical Nutrition*, **44**, 417–24.

Lewis, H.E., Masterton, J.P. & Ferres, H.M. (1958). Selection of representative sites for measuring changes in human subcutaneous tissue thickness. *Clinical Science*, **17**, 369–76.

Lohman, T.G. (1981). Skinfolds and body density and their relation to body fatness: a review. *Human Biology*, **53**, 181–25.

Lohman, T.G., Boileau, R.A. & Massey, B.H. (1975). Prediction of lean body mass in young boys from skinfold thickness and body weight. *Human Biology*, **47**, 245–62.

Lukaski, H.C., Johnson, P.E., Bolonchuk, W.W. & Lykken, G.I. (1985). Assessment of fat-free mass using bioelectric impedance measurements of the human body. *American Journal of Clinical Nutrition*, **41**, 810–7.

Malina, R.M. (1983). Menarche in athletes: a synthesis and hypothesis. *Annals of Human Biology*, **10**, 1–24.

Martin, A.D., Ross, W.D., Drinkwater, D.T. & Clarys, J.P. (1985). Prediction of body density by skinfold caliper: Assumptions and cadaver evidence. *International Journal of Obesity*, **9**, 31–9.

Parizkova, J. (1961). Total body fat and skinfold thickness in children. *Metabolism*, **10**, 794–807.

Parra, A., Argote, R., Garcia, G., Cervantes, C., Alatorre, S. & Perez-Pasten, E. (1979). Body composition in Hypopituitary Dwarfs Before and during Human Growth Hormone Therapy. *Metabolism*, **28**, 851–7.

Preece, M.A., Round, J.M. & Jones, D.A. (1987). Growth hormone deficiency in adults – an indication for therapy. *Acta Paediatrica Scandinavica (suppl.)*, **331**, 76–9.

Satwanti, I., Singh, P. & Bharadwaj, H. (1980). Fat distribution in lean and obese young women: A densitometric and anthropometric evaluation. *American Journal of Physical Anthropology*, **53**, 611–6.

Schoeller, D.A., Van Santan, E., Peterson, D.W., Dietz, W., Jaspan, J. & Klein, P.D. (1980). Total body water measurements in humans with O^{18} and ^2H labeled water. *American Journal of Clinical Nutrition*, **33**, 2686–93.

Sloan, A.W. (1967). Estimation of body fat in young men. *Journal of Applied Physiology*, **23**, 311–5.

Tanner, J.M. & Whitehouse, R.H. (1975). Revised standards for triceps and subscapular skinfolds in British children. *Archives of Disease in Childhood*, **50**, 142–5.

Tanner, J.M., Whitehouse, R.H., Hughes, P.C.R. & Vince, F.P. (1971). Effect of human growth hormone treatment for 1 to 7 years on growth of 100 children, with growth hormone deficiency, low birth weight, inherited smallness, Turners syndrome and other complaints. *Archives of Disease in Childhood*, **46**, 745–82.

Thomasset, A. (1962). Bio-electrical properties of tissue impedance measurements. *Lyon Medical*, **207**, 107–18.

Thomasset, A. (1963). Bio-electrical properties of tissues. *Lyon Medical*, **209**, 1325–52.

Ulin, K., Meydani, M., Zammenhof, R.G. & Blumberg, J.B. (1986). Photon activation analysis as a new technique for body composition. *American Journal of Clinical Nutrition*, **44**, 963–73.

van Loan, M. & Mayclin, P. (1987). A new TOBEC instrument and procedure for the assessment of body composition: use of Fourier coefficients to predict lean body mass and total body water. *American Journal of Clinical Nutrition*, **45**, 131–7.

Young, C.M., Sipin, S.S. & Roe, D.A. (1968). Body composition studies of pre-adolescent and adolescent girls. III. Predicting specific gravity. *Journal of the American Dietetic Association*, **53**, 469–75.

Part III

Growth and tissue factors

DAVID R. CLEMMONS, H. WALKER BUSBY
& LOUIS E. UNDERWOOD

Mediation of the growth promoting actions of growth hormone by somatomedin-C/insulin like growth factor I and its binding protein

Growth hormone (GH) exerts a variety of metabolic actions *in vivo*. These diverse effects, which include alterations in lipid and carbohydrate metabolism as well as the stimulation of growth, are brought about through multiple mechanisms. This chapter will be focused on the interrelationship between GH and somatomedin-C/Insulin like growth factor I (SM-C/IGF-I), a mediator of the growth promoting action of GH. While not intended to be a comprehensive review of either GH or SM-C/IGF-I,we hope to provide the reader with insights into current understanding of the actions of these peptides.

Growth hormone has both direct and indirect actions

GH is secreted by the pituitary in an episodic pulsatile manner, with secretory bursts occurring several times daily (Bercu & Diamond, 1986). Despite these secretory bursts, the stimulation of SM-C/IGF-I secretion by growth hormone requires 4 or more hours to be detected in blood (Copeland *et al.*, 1980). In direct contrast, plasma concentrations SM-C/IGF-I are quite stable, and reflect integrated GH secretion rather than minute to minute fluctuations. Although the precise role of pulsatile GH secretion is unknown, it may be involved in the direct effects of GH on carbohydrate and lipid metabolism since these changes occur very rapidly after GH administration (Goodman & Knobil, 1961) (Figure 1). Injection of GH causes breakdown of triglycerides into free fatty acids and glycerol (Rabel & Hollenberg, 1959), and produces rapid increases in blood glucose. The net result of these direct actions of GH may be to facilitate the utilization of lipid and carbohydrate substrates to satisfy the energy requirements of growing tissues. However, the mechanisms that link these metabolic processes to long-term growth are not defined.

In contrast to these short-term effects, several lines of evidence support the hypothesis that the long-term growth-promoting effects of GH are mediated through SM-C/IGF-I the plasma concentrations of which rise following injection of GH into hypopituitary patients (Copeland *et al.*, 1980). SM-C/IGF-I stimulates sulfate incorporation by cartilage *in vitro*, DNA synthesis by many cell types (Clemmons & Van Wyk, 1981), and protein synthesis in multiple tissues (Salmon & Du Vall, 1970). While GH has little or no direct effect in these *in vitro* test systems, Sm-C/IGF-I administered *in vivo* has been shown to stimulate growth of cartilage and soft tissues (Schoenle *et al.*, 1982).

Controversy has existed for years over whether all of the growth-promoting effects of GH were mediated through SM-C/IGF-I, or whether GH promotes growth directly. Nixon and Green (1984) have contributed to resolving this controversy by showing that GH probably stimulates growth through both direct and indirect mechanisms. Specifically, GH stimulates 3T3 L1 cell precursors to differentiate into adipocytes (Morikawa, Nixon & Green, 1982). GH then stimulates an increase in the concentration of SM-C/IGF-I, and this in turn stimulates division of the cells that have been committed to differentiate (Zezulak & Green, 1986). This hypothesis has been extended by showing that GH not only causes prechondrocytes to commit to a differentiated pathway, but once this event has commenced, GH further stimulates the synthesis of Sm-C/IGF-I by the committed cells (Nilsson *et al.*, 1986). GH seems to exert this effect following direct local injection into cartilage, and blood transport of GH or IGF-I does not appear to be required.

Figure 1. Possible scheme for categorizing the metabolic actions of GH. From Underwood & Van Wyk, (1985).

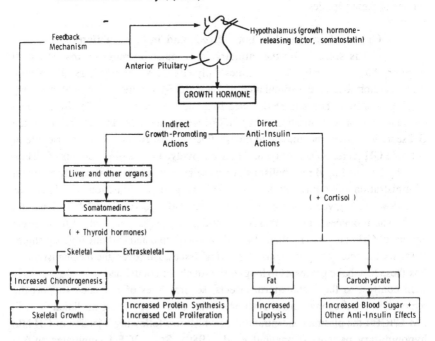

Sites of Sm-C/IGF-I synthesis

Increases in plasma Sm-C/IGF-I concentrations were believed for several years to result from hepatic synthesis of Sm-C/IGF-I and its release into the circulation for transport to distant sites (Van Wyk & Underwood, 1978). The autocrine/paracrine hypothesis of growth factor action (Sporn & Roberts, 1985), however, has stimulated consideration of the possibility that other tissues also might be important sources of Sm-C/IGF-I, and reevaluation of the question of whether blood transport is required for this peptide to stimulate growth. Many investigators have shown evidence that somatomedins are produced by many tissues, including rat liver (Moses *et al.*, 1980a), multiple tissues of fetal mice (D'Ercole, Applewhite & Underwood, 1980) and a variety of cultured cell types (Van Wyk, 1984). We and others have shown that cultured human skin fibroblasts synthesize a form of Sm-C/IGF-I, and that release of this substance can be stimulated by growth hormone (Atkison *et al.*, 1980); Clemmons *et al.*, 1981a). From subsequent studies it has been shown that fibroblasts release a precursor form of Sm-C/IGF-I (Clemmons & Shaw, 1986) and when isolated in pure form this peptide stimulates fibroblast division. C-terminal sequence analysis of this protein has shown it to be identical to the peptide predicted from the analysis of an alternatively processes mRNA reported from human liver (Rotwein, 1986). Furthermore, blocking the binding of this precursor to the type 1 IGF receptor on fibroblasts inhibits fibroblast growth (Clemmons & Van Wyk, 1985). The results of interrupting this local microcircuit suggest that the secreted material stimulates growth by an autocrine or paracrine mechanism. Sm-C/IGF-I has also been shown to be present in extracts of adult rat tissues. The tissue Sm-C/IGF-I concentrations are reduced in hypophysectomized rats and can be increased by injection of GH (D'Ercole, Stiles & Underwood, 1984). Furthermore, increases in tissue Sm-C/IGF-I can be detected prior to increases in the plasma Sm-C/IGF-I concentration. Sm-C/IGF-I mRNA can also be detected in multiple tissues, and induction of an 8–10 fold increase in Sm-C/IGF-I mRNA can be detected in both hepatic and non-hepatic tissues following injection of GH to hypopituitary animals (Hynes *et al.*, 1987). Preliminary data indicate that changes in nutritional intake can alter Sm-C/IGF-I mRNA expression in response to GH administration to hypophysectomized animals (Emler & Schach, 1987). These findings suggest that Sm-C/IGF-I may be secreted locally in many tissues, and stimulate growth prior to its entry into the circulation.

From observations made using *in situ* hybridization techniques, Han and coworkers have concluded that fibroblasts and or fibroblast-like cells in the structural elements of most tissues contain the Sm-C/IGF-I mRNA (Han *et al.*, 1987). In contrast, immunocytochemical localization studies show that the somatomedins are present on the surface of many cell types that do not have hybridizable mRNA, and presumably are not sites on synthesis. Since the 35 KDa IGF binding protein (see below) is also present on the surfaces of these cell types (V.K.M. Han *et al.*, unpublished observations), the immunochemical localization that has been observed may be due to local increases in the binding protein concentration. Likewise this protein could function as a transport protein to certain specific cell types and enhance Sm-C/IGF-I binding to cell surface receptors (Clemmons *et al.*, 1986).

Multiple lines of evidence, therefore, indicate that many tissues synthesize and secrete Sm-C/IGF-I however the precise role of local secretion in stimulating systematic growth remains to be determined. Current data suggest, however, that peptide secreted locally may be as important a growth stimulator as the Sm-C/IGF-I transported in the blood.

IGF receptors

Enhancing the possibility that Sm-C/IGF-I acts by autocrine/paracrine mechanisms is the observation that IGF receptors are ubiquitous, being present on almost all cell types. Two subtypes of IGF receptors have been described: The type I receptor is a heterotetramer composed of two 135 kDa subunits that contain the Sm-C/IGF-I binding site and two 95 dKa subunits that contain tryrosine kinase activity (Massague *et al.*, 1981; Kasuga *et al.*, 1981). Immediately following association of Sm-C/IGF-I with the type I receptor, the tyrosines that are present on the beta subunit are phosphorylated. Recently, it has been reported that other intracellular proteins may be phosphorylated as a direct result of Sm-C/IGF-I association with the type I receptor (Izumi *et al.*, 1987). The order of potency in competing for binding to this receptor is Sm-C/IGF-I>IGF-II>insulin. The type II receptor is a 260 kDa monomer that binds IGF-II>IGF-I and does not bind insulin. Although the relative importance of these two types of receptors in mediating growth responses are not known, in L–6 myoblasts the type I receptor appears to mediate the growth response regardless of whether Sm-C/IGF-I or IGF-II is used as a stimulant (Ewton, Raben & Florini, (1987).

Following the binding of Sm-C/IGF-I to the type I receptor, there is down-regulation of the receptor (Rosenfeld & Hintz, 1980) and this may function as a mechanism of protecting cells from overstimualtion. Kaplowitz has reported recently that exposure of cultured fibroblasts to high concentrations of dexamethasone induces an increase in IGF receptor number (Kaplowitz, 1987). The hormonal milieu therefore may also be important in controlling the cellular response to this growth factor. This may explain why pretreatment of fibroblast cultures with dexamethasone enhances the cellular DNA synthesis response to Sm-C/IGF-I (Conover *et al.*, 1985). Platelet-derived growth factor (PDGF) also has been shown to induce an increase in IGF receptor number in cultured cells, providing a mechanism by which the cellular response to Sm-C/IGF-I could be enhanced (Clemmons, Van Wyk & Pledger, 1980).

Actions of Sm-C/IGF-I *in vitro*

Because limited quantities were available, most studies of Sm-C/IGF-I's actions on target cells have been conducted using *in vitro* test systems. While these studies provide valuable insights regarding the biochemical events that mediate growth at the cellular level, the roles of receptor subtypes in mediating Sm-C/IGF-I actions, and the mechanisms by which Sm-C/IGF-I might stimulate growth, they all suffer from the problems of potential changes of responses during long-term cell culture and of trying to mimic *in vivo* conditions in *in vitro* test systems. Under *in vitro* conditions, factors such as extracellular matrix, cell-cell interactions, hormonal milieu and nutrient status may differ from those encountered *in vivo*. Despite these limitations, much has been learned about the actions of Sm-C/IGF-I.

In the original description of what we now recognize as a Sm-C/IGF-I effect, an active principal of serum was shown to stimulate proteoglycan synthesis by cartilage and thymidine incorporation into cartilage DNA (Salmon & Daughaday, 1957). This growth promoting property has now been extended to many different cell types. Sm-C/IGF-I appears to act in part by potentiating the effect of other growth factors. When added alone *in vitro* it has modest effects on cell growth, but when added with other mitogens such as PDGF or epidermal growth factor (EGF), it induces synergistic increases in DNA synthesis (Leof *et al.*, 1982). In mouse fibroblasts, Sm-C/IGF-I acts in the latter part of the G1 phase of the cell cycle, but has no effect on quiescent cells (Stiles *et al.*, 1979). Therefore, mitogens such as PDGF are required to activate cells to enter the cell cycle in order for Sm-C/IGF-I to be effective. Sm-C/IGF-I has been shown to stimulate increases in the protein content of cells and in cell size. In specific target cell systems Sm-C/IGF-I potentiates the expression of differentiated cell functions, such as induction of creatine kinase in L6 myoblasts (Ewton & Florini, 1981), and potentiation of FSH-stimulated steriodogenesis by granulosa cells (Adashi *et al.*, 1985) (Figure 2).

Mechanisms of Sm-C/IGF-I actions
Role of binding proteins

The mechanisms by which Sm-C/IGF-I modulates the growth of target cells have been the focus of intense study. Following secretion into extracellular fluids, Sm-C/IGF-I associates with 32–35 kDa binding proteins (Chochinov *et al.*, 1977). While the existence of these proteins is not in doubt, their chemical structures have not been defined fully. Using assays that measure binding of radiolabelled Sm-C/IGF-I or IGF-II, these proteins have been shown to be present in spinal fluid (Zapf *et al.*, 1984), amniotic fluid (Drop *et al.*, 1979), lymph (Moses *et al.*, 1979), blood (Drop *et al.*, 1984), human luteal fluid (Seppala *et al* , 1984) and in supernatants of many types of cultured cells (Nissley *et al.*, 1977). Binoux *et al.* (1982) have presented evidence that at least two forms of these binding proteins are present in spinal fluid based on their affinities for Sm-C/IGF-I and IGF-II. Likewise, investigators have presented structural data that are consistent with the existence of multiple forms (Hossenlopp *et al.*, 1986). Plasma also contains a larger (e.g., 57 K) IGF binding protein. This protein is immunochemically distinct from the extracellular fluid forms, is secreted by hepatocytes (Scott *et al.*, 1985) and its plasma concentration is growth hormone dependent (Wilkins & D'Ercole, 1985).

Detailed structural analyses of one form of these binding proteins has been completed by Povoa *et al.* (1984). After purification from human amniotic fluid and from human placenta, this protein was shown to have a Mr of 34,000 daltons and an isoelectric point of 4.8. It binds one molecule of Sm-C/IGF-I per molecule of binding protein, and impure preparations inhibit Sm-C/IGF-I-stimulated cartilage sulfation (Drop *et al.*, 1984). Using an RIA, its concentrations in plasma were observed to be related inversely to growth hormone secretory status (Drop *et al.*, 1984). Immunoblotting has confirmed its presence in extracellular fluids and culture supernatants (Hossenlopp *et al.*, 1986). The factors controlling synthesis of this binding protein are not well defined; however, estrogen has been shown to induce its synthesis by decidual explants (Rutanen *et al.*, 1986). Plasma concentrations of the protein are elevated late in pregnancy and in fetal and cord blood (Povoa *et al.*, 1984).

Figure 2. Synergism between Sm-C/IGF-I and FSH in stimulating progesterone accumulation by cultured rat granulosa cells. Granulosa cells from immature hypophysectomized diethylstilbesterol-treated rats were cultured (1×10^5/dish) for 72 hours without serum. The concentration of FSH added is 20 ng/ml. Data points represent mean ± SE, n = 4. From Adashi *et al.*, 1984.

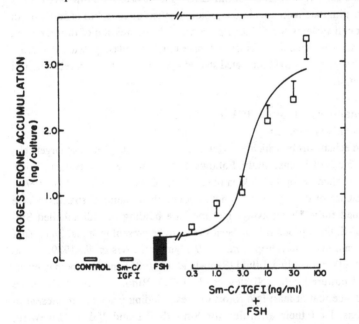

Its concentration declines rapidly after birth, and is inversely related to growth rate in early life (Drop *et al.*, 1984).

Our laboratory has shown that two forms of the IGF binding protein are present in human amniotic fluid (Clemmons *et al.*, 1987). One form adheres to cell surfaces, whereas another which has an identical amino acid composition does not adhere. The cell surface adherent form potentiates cellular DNA synthesis in response to Sm-C/IGF-I, and the combination of Sm-C/IGF-I and the binding protein increases DNA synthesis to levels that were comparable to those produced by 10% serum (Elgin *et al.*, 1987). In contrast, the form of the Sm-C/IGF-I binding protein that does not adhere to cell surfaces inhibits fibroblast DNA synthesis in response to Sm-C/IGF-I (D.R. Clemmons *et al.*, unpublished observation). The distribution of these two forms of binding proteins in various extracellular fluids is now being determined, as is the chemical property that accounts for membrane adherence. Human fibroblasts secrete the membrane adherent form of the protein and it can be detected on the surface of fibroblasts and in the culture medium (Clemmons *et al.*, 1986). When binding proteins are present on the surface of the fibroblast, the cells repond to low concentrations of Sm-C/IGF-I by increasing their synthesis of DNA, even if there are no other growth factors present in the medium. Furthermore, blocking of Sm-C/IGF-I binding to the 34 kDa protein on the cell surface with antibody to the 34 kDa protein results in a significant diminution of Sm-C/IGF-I-induced DNA synthesis. One mechanism by which this protein might enhance the cellular response to Sm-C/IGF-I was suggested by showing that when the binding protein is adherent to fibroblast surfaces, binding of radiolabelled Sm-C/IGF-I to the type I receptor is paradoxically increased (Clemmons *et al.*, 1986). This potentiation of Sm-C/IGF-I binding by the binding protein suggests that the latter may function by altering the affinity of the type I receptor for Sm-C/IGF-I. The binding protein, however, might also alter the cellular response to Sm-C/IGF-I directly. If Sm-C/IGF-I binding to the type 1 receptor is blocked using a specific anti-receptor antibody, the combination of Sm-C/IGF-I plus the binding protein is sufficient to enhance the cellular response to this growth factor, although the degree of increase is less than that seen using cells not exposed to anti-receptor antibody. The mechanism by which the binding protein might transmit the Sm-C/IGF-I mitogenic signal independent of the receptor has not been determined.

Factors controlling Sm-C/IGF-I secretion
Age and hormones

Plasma Sm-C/IGF-I concentrations are low in newborns and increase progressively from a mean of 0.31 ± 0.11 U/ml at 1 year to more than 2.5 U/ml at puberty (Clemmons & Van Wyk, 1983). Values are 20% higher in females than males before, during and after puberty (Figure 3). From the pubertal peak that corresponds to Tanner stage III for pubic hair, breasts, or genitalia, there is a significant decline of values to approximately 2.0 U/ml at 20 years, then a slower decline to values of 0.5–0.7 u/ml in the seventh decade of life. The factors controlling the increases prior to puberty are unknown, as are those that produce the age-related decline. The pubertal increases in plasma Sm-C/IGF-I concentrations may result from sex steroid-induced increases in GH secretion (Zadik *et al.*, 1985). It is clear, however, that androgens have no direct effect on Sm-C/IGF-I secretion (Craft & Underwood, 1984). Likewise estrogens in physiologic concentrations appear to increase GH secretion (Cutler *et al.*, 1985), although administration of pharmacologic amount of estrogen may inhibit Sm-C/IGF-I secretion directly (Clemmons *et al.*, 1980).

Hormones such as thyroxine and cortisol may modulate Sm-C/IGF-I secretion or action. Hypothyroid patients often have low plasma Sm-C/IGF-I concentrations, although their rise in Sm-C/IGF-I after injection of GH is normal, suggesting that the effect of T_4 may be mediated through GH secretion (Chernausek *et al.*, 1983). While cortisol appears to attenuate the response of target cells to Sm-C/IGF-I, it does not affect blood concentrations of the hormone (Gourmelen, Girard & Binoux, 1982).

Figure 3. Plasma Sm-C/IGF-I concentrations during childhood and adolescence. The values in normal children were determined on EDTA plasma samples from 846 children. The vertical bars indicate the mean ± 95% confidence limits for normal boys (hatched bars) and girls (open bars). The circles are values on iandividual children with severe GH deficiency. Reproduced from Underwood & Van Wyk, (1985).

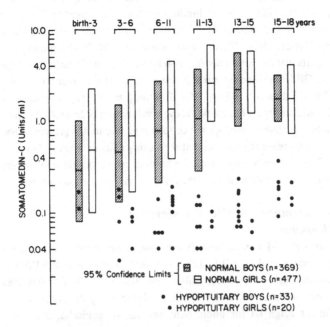

Control by growth hormone

Sm-C/IGF-I concentrations appear to reflect GH secretory rates in patients with hypopituitarism or treated acromegaly. Thus children who have subnormal 24 hour GH secretion but who are not completely GH deficient have Sm-C/IGF-I values that are lower than normal (Bercu *et al.*, 1986). Likewise children wo have no detectable GH secretion after provocative stimuli have Sm-C/IGF-I values that are lower than those with subnormal GH secretion. In contrast, patients with acromegaly have elevated concentrations of Sm-C/IGF-I (Clemmons *et al.*, 1979). The highest Sm-C/IGF-I values are observed in patients with untreated acromegaly and more modest elevations occur in partially-treated patients. Several investigators have shown that the magnitude of the elevation in Sm-C/IGF-I parallels the severity of the abnormality of GH secretion in these partially-treated patients (Clemmons *et al.*, 1979; Rieu *et al.*, 1982; Chiodini *et al.*, 1987).

Nutrition and the regulation of Sm-C/IGF-I

One of the principal regulators of Sm-C/IGF-I concentrations in plasma is nutritional status. In a study done in our clinic 7 obese adult subjects who were fasted for 10 days experienced a 70% decline in plasma Sm-C/IGF-I concentrations (Clemmons *et al.*, 1981b). The fasting induced decline in Sm-C/IGF-I may be related to GH resistance since Merimee *et al.* (1984) showed that three days of fasting is sufficient to obliterate the plasma Sm-C/IGF-I response to GH injections.

We have also performed studies attempting to define the dietary variables that are required to restore Sm-C/IGF-I after fasting. We have observed that following a 5 day fast, ingestion of a threshold amount of energy (between 11–18 kcal/kg) is required for normal volunteers to have a detectable increment in plasma Sm-C/IGF-I concentration (Isley, Underwood & Clemmons, 1984). In contrast, if sufficient energy is ingested any amount of protein in the refeeding diet produces increments in plasma Sm-C/IGF-I concentration. Normal prefast Sm-C/IGF-I concentrations, however, are not reached unless 1.0 gm protein/kg is ingested. In several studies, we have observed that the rate at which Sm-C/IGF-I secretion is restored after fasting is slower than the rate of decline during fasting, indicating that the restoration of this peptide to normal level may require optimum ingestion of several dietary components. One of these components may be essential amino acids, because supplementation with essential amino acids accelerates the return of plasma Sm-C/IGF-I toward normal in subjects fed a nitrogen deficient diet after fasting (Clemmons, Seek & Underwood, 1985).

We have observed repeatedly that diet-induced changes in plasma Sm-C/IGF-I correlate well with changes in nitrogen balance during refeeding (Isley, Underwood & Clemmons, 1983). Since Sm-C/IGF-I has direct stimulatory effects on protein synthesis in muscle, it is possible that diet-induced changes in Sm-C/IGF-I concentrations are directly linked to changes in nitrogen conservation. Measurement of Sm-C/IGF-I, therefore, may be a useful indicator of the response to nutritional support in patients undergoing hyperalimentation therapy. In this regard, we have reported a study in which six malnourished subjects who were 8–15% below ideal body weight were given 1.3 gm protein/kg and 39–42 kcal/kg for 16 days (Clemmons *et al.*, 1985). Plasma Sm-C/IGF-I concentrations rose from a mean of 0.67 ± 0.23 U/ml (ISD) to 2.44 U/ml by day 10 of therapy. Subsequently the mean plasma Sm-C/IGF-I reached a new plateau at 1.51 ± 0.37 U/ml by day 16. During the first ten days the plasma Sm-C/IGF-I response paralleled the changes that occurred in the nitrogen balance. However, between days 10–16, nitrogen balance remained positive (albeit less so than previously) at a time when plasma Sm-C/IGF-I concentrations were falling, indicating that the change in Sm-C/IGF-I may be more sensitive to acute directional changes in nitrogen balance than to absolute balance. Interestingly, the rate of protein turnover has also been shown to decline during the period corresponding to days 10–16 of refeeding (Golden, Waterlow & Picou, 1979). Therefore, it is possible that changes in plasms Sm-C/IGF-I are more closely related to this parameter.

Interaction between GH and nutrition

We have studied obese volunteers to assess the interaction between GH and nutrition, because such subjects are often willing to undergo chronic caloric restriction. Reduction of caloric intake to 12 kcal/kg in subjects who are 30–70% over ideal body weight results in a significant decline in plasma Sm-C/IGF-I concentrations to 78% of control by 12 days (Williams, Underwood & Clemmons, unpublished observations). The decline in Sm-C/IGF-I under such conditions, however, is less than that observed in normal weight volunteers. Furthermore, no decline occurs when obese volunteers are restricted to 18 kcal/kg, whereas normal weight subjects that are fed this diet have a decrease of Sm-C/IGF-I to 54% of control by 5 days. Obese subjects, therefore, appear to require fewer calories in order to maintain normal blood Sm-C/IGF-I concentrations and more severe restriction to effect significant decreases.

Administration of GH exogenously to obese subjects, even when they are calorically restricted, results in an increase in plasma Sm-C/IGF-I concentrations (Clemmons *et al.*, 1987b) (Figure 4). When obese subjects were fed a diet that contained a 33% reduction from optimal caloric intake (24 kcal/kg) and GH given at a dose of 0.1 mg/kg every other day, plasma Sm-C/IGF-I rose promptly to 3.2 times the values observed during injection of vehicle, and remained elevated during 3 weeks of GH treatment. Likewise, nitrogen balance while receiving GH was 2.6 gms/day more positive than during injection of vehicle. We also have observed a similar increase in plasma Sm-C/IGF-I in subjects whose caloric intake was restricted to 18 kcal/kg.

We conclude, therefore, that more severe caloric restriction will be needed to attenuate the usual Sm-C/IGF-I response to GH given exogenously, and that the degree of caloric restriction at which the transition to GH unresponsiveness occurs in obese subjects is probably different from normal individuals (Snyder *et al.*, 1987, unpublished observation). During the study in which 18 kcal/kg ideal body weight was ingested, plasma Sm-C/IGF-I concentrations rose to maximal values (4.3 fold over control) after 9 days of therapy (0.1 mg/kg), and remained increased for the 11 weeks of therapy. Nitrogen conservation in these subjects was excellent during the first 5 weeks. In contrast, nitrogen conservation was attenuated of after 6 weeks of GH, and was not significantly different from subjects given injections of vehicle. Similarly, the increment in plasma free fatty acids in response to GH, which was excellent for the first 6 weeks, but declined in the latter 5 weeks of GH therapy

Figure 4. The influence of growth hormone (0.1 mg/kg given every other day) on nitrogen balance and plasma Sm-C/IGF-I. From Clemmons *et al.* (1987b).

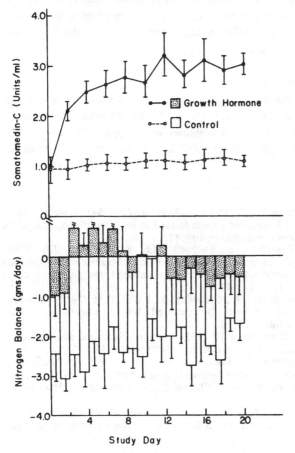

(increment was 0.78 meq/l before the 6th week and 0.23 meq/l after 6 weeks). These observations suggest that the study subjects had become refractory to the nitrogen sparing effects of Sm-C/IGF-I, and that the direct effects of GH on lipolysis had been reduced. The mechanism(s) accounting for these phenomena are undefined. It is possible that they are adaptive mechanisms to semi-starvation and that resistance to GH and/or Sm-C/IGF-I is a component of the adaptive response. These observations point to the need to evaluate the effect of nutritional status on GH responsiveness during long-term therapy.

Summary

Much progress has been made in understanding the structure and biologic properties of Sm-C/IGF-I. Major gaps in our knowledge include understanding of the factors that control secretion and proteolytic processing of the molecule at the cellular level. Likewise the interaction between nutritional status, growth hormone, and other hormones in regulating Sm-C/IGF-I secretion has been incompletely defined. The role of cell-secreted Sm-C/IGF-I in growth and of the binding proteins in controlling the local concentrations of Sm-C/IGF-I needs further study, as do the intracellular factors that mediate these responses.

Acknowledgements

The authors gratefully acknowledge the technical assistance of Eyvonne Bruton, Karen Koerber and Mary George Johnson. We thank Emilia Richichi for her help in preparing the manuscript. These studies were supported by Grants AG–02331, HL–36311, HD–08299 and AM–01022 from the National Institutes of Health.

References

Adashi, E.Y., Resnick, CE., Svoboda, M.E. *et al.*, (1984). A novel role for somatomedin-C/insulin-like growth factor I in the cytodifferentiation of the ovarian granulosa cell. *Endocrinology*, **115**, 1227–9.

Adashi, E.Y., Resnick, C.E., Svoboda, M.E. & Van Wyk, J.J. (1985). Somatomedin-C synergizes with follicle-stimulating hormone in the acquisition of progestin biosynthetic capacity by cultured rat granulosa cells. *Endocrinology*, **116**, 2135–42.

Atkison, P.R., Weidman, E.R., Bhaumick, B. & Bala, R.M. (1980). Release of somatomedin-like activity by cultured WI–38 human fibroblasts. *Endocrinology*, **106**, 2006–12.

Bercu, B.B. & Diamond, F.B. (1986). Growth hormone neurosecretory dysfunction. *Clinics in Endocrinology and Metabolism*, **15**, 537–77.

Bercu, B.B., Shulman, D., Root, A.W. & Spiliotis, B.S. (1986). Growth hormone provocative testing frequently does not reflect endogenous growth hormone secretion. *Journal of Clinical Endocrinology and Metabolism*, **63**, 768–73.

Binoux, M., Hardouin, S., Lassare, C. & Hossenlopp, P. (1982). Evidence for production by the liver of two IGF binding proteins with similar molecular weights but different affinities for IGF-I and IGF-II. Their relationship with serum and cerebrospinal fluid binding proteins. *Journal of Clinical Endocrinology and Metabolism*, **55**, 600–2.

Chernausek, S.D., Underwood, L.E., Utiger, R.D. & Van Wyk, J.J. (1983). Growth hormone secretion and plasma somatomedin-C in primary hypothyroidism. *Clinical Endocrinology*, **19**, 334–7.

Chiodini, P.G., Cozzi, D., Dallabonzana, G., Opizzi, G., Verde, M., Petrochini, M., Luzzi, A. & Del Pozo, E. (1987). Medical treatment of acromegaly with SMS 201–995, a somatostatin analog: A comparison with bromocryptine. *Journal of Clinical Endocrinology and Metabolism*, **64**, 447–53.

Chochinov, R.H., Mariz, I.K., Hajek, A.S. & Daughaday, W.H. (1977). Characterization of a protein in mid-term human amniotic fluid that reacts in the somatomedin-C radioreceptor assay. *Journal of Clinical Endocrinology and Metabolism*, **44**, 902–8.

Clemmons, D.R., Van Wyk, J.J., Ridgway, E.C., Kliman, B., Kjelberg, R.N. & Underwood, L.E. (1979). Evaluation of acromegaly by radioimmunoassay of somatomedin-C. *New England Journal of Medicine*, **301**, 1138–43.

Clemmons, D.R., Van Wyk, J.J. & Pledger, W.J. (1980). Sequential addition of platelet factor and plasma to Balb/c 3T3 fibroblast cultures stimulates somatomedin-C binding early in the cell cycle. *Proceedings of the National Academy of Sciences, USA*, **77**, 6644–8.

Clemmons, D.R., Underwood, L.E., Ridgway, E.C., Kliman, B. & Van Wyk, J.J. (1980). Estradiol treatment of acromegaly: reduction of immunoreactive somatomedin-C and improvement of metabolic status. *American Journal of Medicine*, **69**, 571–5.

Clemmons, D.R. & Van Wyk, J.J. (1981). Somatomedin: physiological control and effects on cell proliferation. In *Handbook of Experimental Pharmacology*, vol. **57**, ed. R. Baserga, pp. 161–208. Springer-Verlag, Berlin.

Clemmons, D.R., Underwood, L.E. & Van Wyk, J.J. (1981a). Hormonal control of immunoreactive somatomedin production by cultured human fibroblasts. *Journal of Clinical Investigation*, **67**, 10–19.

Clemmons, D.R., Kibanski, A., Underwood, L.E., McArthur, J.W., Ridgway, E.C., Beitens, I.Z. & Van Wyk, J.J. (1981b). Reduction in immunoreactive somatomedin-C during fasting. *Journal of Clinical Endocrinology and Metabolism*, **53**, 1247–50.

Clemmons, D.R., Van Wyk, J.J. (1983). Factors controlling blood concentration of somatomedin-C. *Clinics in Endocrinology and Metabolism*, **13**, 113–43.

Clemmons, D.R., Van Wyk, J.J. (1985). Evidence for a functional role of endogenously produced somatomedin-like peptide in the regulation of DNA synthesis in cultured human fibroblasts and porcine smooth muscle cells. *Journal of Clinical Investigation*, **75**, 1914–8.

Clemmons, D.R., Seek, M.M. & Underwood, L.E. (1985). Supplemental essential amino acids augment the somatomedin-C/insulin-like growth factor I response to refeeding after fasting. *Metabolism*, **34**, 391–5.

Clemmons, D.R., Underwood, L.E., Dickerson, R.N. *et al.* (1985). Use of somatomedin-C/insulin-like growth factor I measurements to monitor the response to nutritional repletion in malnourished patients. *American Journal of Clinical Nutrition*, **41**, 191–8.

Clemmons, D.R. & Shaw, D.S. (1986). Purification and biologic properties of fibroblast somatomedin. *Journal of Biological Chemistry*, **261**, 10293–8.

Clemmons, D.R., Elgin, R.G., Han, V.K.M., Casella, S.J., D'Ercole, A.J. & Van Wyk, J.J. (1986). Cultured fibroblast monolayers secrete a protein that alters the cellular binding of somatomedin-C/insulin-like growth factor I. *Journal of Clinical Investigation*, **77**, 1548–58.

Clemmons, D.R., Williams, R., Snyder, D.K. & Underwood, L. (1987b). Treatment with growth hormone conserves lean body mass during dietary restriction in obese volunteers. *Journal of Clinical Endocrinology and Metabolism*, **64**, 878–83.

Copeland, K.C., Underwood, L.E. & Van Wyk, J.J. (1980). Induction of immunoreactive somatomedin-C in human serum by growth hormone: dose response relationships and effect on chromatographic profiles. *Journal of Clinical Endocrinology and Metabolism*, **50**, 690–7.

Conover, C.A., Dollar, L.A., Hintz, R.L. & Rosenfeld, R.G. (1985). Insulin like growth factor I/somatomedin-C synergistically regulates mitosis in competent fibroblasts. *Journal of Cellular Physiology*, B116, 191–8.

Craft, H. & Underwood, L.E. (1984). Effect of androgens on plasma somatomedin-C/insulin-like growth factor I responses to growth hormone. *Clinical Endocrinology*, **20**, 549–54.

Cuttler, L., Van Vliet, G., Conte, P.A. Kaplan, S.L. & Grmbach, M.M. (1985). Somatomedin-C levels in children and adolescents with gonadal dysgenesis: differences from age-matched normal females and effect of chronic estrogen replacement therapy. *Journal of Clinical Endocrinology and Metabolism*, **60**, 1087–92.

D'Ercole, A.J., Applewhite, G.T. & Underwood, L.E. (1980). Evidence that somatomedin is synthesized by multiple tissues in the fetus. *Developmental Biology*, **75**, 315–28.

D'Ercole, A.J., Stiles, A.D. & Underwood, L.E. (1984). Tissue concentrations of somatomedin-C: further evidence for multiple sites of synthesis and paracrine/autocrine mechanisms of action. *Proceedings of the National Academy of Sciences, USA*, **81**, 935–9.

Drop, S.L.S., Valiquette, G., Guyda, H.J., Corvol, M.T. & Posner, B.I. (1979). Partial purification and characterization of a binding protein for insulin-like activity in human amniotic fluid. A possible inhibitor of insulin-like activity. *Acta Endocrinologia*, **90**, 505–18.

Drop, L.S., Kortleve, D.J., Guyda, H.J. & Posner, B.I. (1984). Immunoassay of a somatomedin-binding protein from human amniotic fluid: levels in fetal, neonatal and adult sera. *Journal of Clinical Endocrinology and Metabolism*, **59**, 908–15.

Elgin, R.G., Busby, W.H. & Clemmons, D.R. (1987). An insulin-like growth factor binding protein enhances the biologic response to IGF-I. *Proceedings of the National Academy of Sciences, USA*, **84**, 3313–8.

Emler, C.A. & Schalch, D.S. (1987). Nutritionally-induced changes in hepatic insulin-like growth factor I (IGF-I) gene expression in rats. *Endocrinology*, **120**, 832–4.

Ewton, D.Z. & Florini, J.R. (1981). Effects of somatomedins and insulin on myoblast differentiation in vitro. *Developmental Biology*, **86**, 31–9.

Ewton, D.Z., Falen, S.L. & Florini, J.R. (1987). The type II insulin-like growth factor (IGF) receptor has low affinity for IGF-I analogs: Pleotypic actions of IGFs on myoblasts are apparently mediated by the type I receptor. *Endocrinology*, **120**, 115–23.

Golden, M., Waterlow, J.R. & Picou, D. (1979). The relationship between dietary intake, weight gain, nitrogen balance and protein turnover in man. *American Journal of Clinical Nutrition*, **30**, 1245–52.

Gourmelen, M., Girard, F. & Binoux, M. (1982). Serum somatomedin/insulin-like factor (IGF) and IGF carrier levels in patients with Cushing's syndrome or receiving glucocorticoid therapy. *Journal of Clinical Endocrinology and Metabolism*, **54**, 884–91.

Goodman, H.M. & Knobil, E. (1961). Growth hormone and fatty acid mobilization: the role of the pituitary, adrenal and thyroid. *Endocrinology*, **69**, 187–96.

Han, V.K.M., D'Ercole, A.J. & Lund, P.. (1987). Cellular location of somatomedin (insulin-like growth factor) messenger RNA in the human fetus. *Science*, **236**, 193–7.

Hossenlopp, P., Seurin, D., Sevogia-Quinson, B. & Binoux, M. (1986). Identification of an insulin-like growth factor binding proteins in human spinal fluid with selective affinity for IGF-II. *FEBS Letters*, **208**, 439–44.

Hynes, M.A., Van Wyk, J.J., Brooks, P.J., D'Ercole, A.J., Jansen, M.O. & Lund, P.K. (1987). Growth hormone dependence of somatomedin-C/insulin-like growth factor I and insulin-like growth factor II messenger ribonucleic acids. *Molecular Endocrinology*, **1**, 233–42.

Isley, W.L., Underwood, L.E. & Clemmons, D.R. (1983). Dietary components that regulate serum somatomedin-C concentrations in humans. *Journal of Clinical Investigation*, **71**, 175–82.

Isley, W.L., Underwood, L.E. & Clemmons, D.R. (1984). Change in plasma somatomedin-C in response to ingestion of diets with variable protein and energy content. *Journal of Parenteral and Enteral Nutrition*, **8**, 407–11.

Izumi, T., White, M.F., Takashi, K., Takaku, F., Akanuma, Y. & Kusauga, M. (1987). Insulin like growth factor I rapidly stimulates tyrosine phosphorylation of a Mr185000 protein in intact cells. *Journal of Biological Chemistry*, **262**, 1282–7.

Kaplowitz, P. (1987). Glucocorticoids enhance somatomedin-C binding and stimulation of amino acid uptake in human fibroblasts. *Journal of Clinical Endocrinology and Metabolism*, **64**, 563–71.

Kasuga, M., Van Obberghen, E., Nissley, S.P. & Rechler, M.M. (1981). Demonstration of two subtypes of insulin-like growth ractor receptors by affinity cross-linking. *Journal of Biological Chemistry*,, **256**, 5305–8.

Koisitinen, R., Kalkkinen, N., Huhtala, M.-L., Seppala, M., Bohn, H. & Rutanen, E.-M. (1986). Placental protein 12 is a decidual protein that binds somatomedin and has an identical N-terminal amino acid sequence with somatomedin binding protein from amniotic fluid. *Endocrinology*, **118**, 1375–8.

Leof, E.B., Wharton, W. Van Wyk, J.J. & Pledger, W.J. (1982). Epidermal growth factor (EGF) and somatomedin-C regulate G$_I$ progression in competent Balb/c–3T3 cells. *Experimental Cell Research*, **141**, 107–15.

Massague, J., Guilette, B.S. & Czech, M.P. (1981). Affinity labelling of multiplication stimulating activity receptors in membranes from rat and human tissues. *Journal of Biological Chemicstry*, **256**, 2122–5.

Merimee, T.J., Zapf, J. & Froesch, E.R. (1982). Insulin-like growth factors in the fed and fasted states. *Journal of Clinical Endocrinology and Metabolism*, **55**, 999–1002.

Morikawa, M., Nixon, E. & Greene, H. (1982). Growth hormone and the adipose conversion of 3T3 cells. *Cell*, **29**, 783–9.

Moses, A.C., Nissley, S.P., Passamani, J., White, R.M. & Rechler, M.M. (1979). Further characterization of growth hormone dependent somatomedin binding proteins produced by rat liver cells in culture. *Endocrinology*, **104**, 536–46.

Moses, A.C., Nissley, S.P., Short, P.A. & Rechler, M.M. (1980a). Purification and characterization of multiplication stimulating activity, insulin-like growth factors purified from rat liver cell conditioned medium. *European Journal of Biochemistry*, **103**, 387–400.

Moses, A.C., Nissley, S.P., Short, P.A., Rechler, M.M., White, A.B. & Higa, O.Z. (1980b). Elevated levels of multiplication stimulating activity, an insulin-like growth factor in fetal rat serum. *Proceedings of the National Academy of Sciences, USA*, **77**, 3649–53.

Nilsson, A., Isgaard, J., Lindahl, A., Dahlstrom, A., Skottner, A. & Isaksson, O.G. (1986). Regulation by growth hormone of number of chondrocytes containing IGF-I in rat growth plate. *Science*, **233**, 571–4.

Nissley, S.P., Rechler, M.M., Moses, A.C., Short, P.A. & Podskalny, J.M. (1977). Proinsulin binds to a growth peptide receptor and stimulates DNA synthesis in chick embryo fibroblasts. *Endocrinology*, **101**, 708–16.

Nixon, B.T. & Green, H. (1984). Growth hormone promotes differentiation of myoblasts and preadipocytes generated by azacytidine treatment of 10T 1/2 cells. *Proceedings of the National Academy of Sciences, USA*, **81**, 3429–32.

Povoa, G., Enberg, G., Jornvall, H. & Hall, K. (1984). Isolation and characterization of a somatomedin binding protein from mid term human amniotic fluid. *European Journal of Biochemistry*, **144**, 199–204.

Raben, M.S. & Hollenberg, C.H. (1959). Effect of growth hormone plasma fatty acids. *Journal of Clinical Investigation*, **38**, 484–90.

Rieu, M., Girard, F., Bricaire, H.C. & Binoux, M. (1982). The importance of insulin-like growth factor (somatomedin) measurements in the diagnosis and surveillance of acromegaly. *Journal of Clinical Endocrinoalogy and Metabolism*, **55**, 147–53.

Rosenfeld, R.G. & Hintz, R.L. (1980). Characterization for a specific receptor for somatomedin-C (Sm-C) on cultured human lymphocytes: evidence that Sm-C modulates homologous receptor concentration. *Endocrinology*, **107**, 1841–8.

Rotwein, P. (1986). Two insulin-like growth factor I messenger RNAs are expressed in human liver. *Proceedings of the National Academy of Sciences, USA*, **83**, 77–81.

Salmon, W.D. Jr. & Daughaday, W.H. (1957). A hormonally controlled serum factor which stimulates sulfate incorporation by cartilage in vitro. *Journal of Laboratory and Clinical Medicine*, **49**, 825–36.

Salmon, W.D. & Du Vall, M.R. (1970). In vitro stimulation of leucine incorporation into muscle and cartilage protein by a serum fraction with sulfation factor activity: differentiation of effects from those of growth hormone and insulin *Endocrinology*, **78**, 1168–74.

Schoenle, E., Zapf, J., Humbel, R.E. & Froesch, E.R. (1982). Insulin-like growth factor I stimulates growth in hypophysectomized rats. *Nature*, **296**, 252–53.

Scott, C.D., Martin, J.L. & Baxter, R.C. (1985). Rat hepatocytes secrete insulin-like growth factor I and binding protein: effect of growth hormone *in vitro* and *in vivo*. *Endocrinology*, **116**, 1102–7.

Seppala, M., Wahlstrom, T., Koskimies, A.I., Tehuunen, A., Rutanen, E.M., Loistinen, R., Huhtaniemi, I., Bohn, H. & Stenman, U.K. (1984). Human preovulatory follicular fluid luteinized cells of hyperstimulated preovulatory follicles and corpus leuteum contain placental protein 12. *Journal of Clinical Endocrinology*, **58**, 505–10.

Sporn, M.B. & Robert, A.B. (1985). Autocrine growth factors and cancer. *Nature*, **313**, 745–7.

Stiles, C.D., Capone, G.T., Scher, C.D. *et al.* (1979). Dual control of cell growth by somatomedin and platelet derived growth factor. *Proceedings of the National Academy of Sciences, USA*, **76**, 1279–83.

Underwood, L.E. & Van Wyk, J.J. (1985), Normal and abberrant growth. In *Williams Textbook of Endocrinology*, ed. J.D. Wilson & D.M. Foster, pp. 155–205. W.B. Saunders, Philadelphia.

Van Wyk, J.J. (1984). The somatomedins: biological actions and physiologic control mechanisms. In *Hormonal Proteins and Peptides*, ed. C.H. Li, vol. 12, 81–125. Academic Press, Orlando.

Van Wyk, J.J. & Underwood, L.E. (1978). The somatomedins and their actions. In *Biochemical Actions of Hormones*, ed. G. Litwack, vol. 5, 101–47 Academic Press, New York.

Wilkins, J.R. & D'Ercole, A.J. (1985). Affinity labeled plasma somatomedin-C/insulin-like growth factor I binding proteins: evidence of growth hormone dependence and subunit structure. *Journal of Clinical Investigation*, 75, 1350–8.

Zadik, Z., Chalen, S.A., McCarter, J.J., Meistas, M. & Kowarski, A.A. (1985). The influence of age on the 24 hour integrated growth hormone in normal individuals. *Journal of Clinical Endocrinology and Metabolism*, 60, 513–4.

Zapf, J., Schmid, C.H. & Froesch, E.R. (1984). Biological and immunological properties of insulin-like growth factors (IGF) I and II. *Clinics in Endocrinology and Metabolism*, 13(1), 3–30.

Zezulak, K.M. & Green, H. (1986). The generation of insulin-like growth factor I sensitive cells by growth hormone action. *Science*, 235, 513–4.

P. DE PAGTER-HOLTHUIZEN, M.JANSEN,
W. BOVENBERG, J.L. VAN DEN BRANDE &
J.S. SUSSENBACH

Somatomedin gene structure and expression

Introduction*

The somatomedins or insulin-like growth factors (IGF) are small polypeptides, which play an important role in fetal and postnatal growth and development (Daughaday et al., 1972; Clemmons & Van Wyk, 1981). The somatomedins circulate in plasma and are bound to specific carrier proteins (Hintz, 1984). The liver is known to be the major site of IGF production; in addition, synthesis has been shown to occur in many other tissues (D'Ercole, Applewhite & Underwood, 1980). The primary structure of two major human IGF peptides has been fully established (Rinderkneckt & Humbel, 1978a; Rinderkneckt & Humbel, 1978b).

IGF-I is a basic peptide of 70 amino acids which is required for growth in postnatal life. For IGF-II, a 67 amino acid neutral peptide, the function is less clear; it may serve an analogous role in fetal development. The nucleotide sequences of complementary DNAs (cDNAs) encoding human IGF-I and IGF-II have been reported (Jansen et al., 1983; Bell et al., 1984; Jansen et al., 1985; Le Bouc et al., 1986; Rotwein, 1986).

From the cDNA characterization it can be deduced that both growth factors are synthesized as larger precursor molecules which undergo extensive processing. Using the cDNAs as specific probes, the chromosomal assignment of the IGF genes has been determined. Both IGFs are encoded by single copy genes; the IGF-I gene maps to the long arm of chromosome 12 and the gene for IGF-II is located on the tip of the short arm of chromosome 11, only 1.4 kilobases (kb) downstream from the insulin gene) Höppener et al., 1985; de Pagter-Holthuizen et al., 1985; Bell et al., 1985; de Pagter-Holthuizen et al., 1987).

In the present communication the organization and expression of the genes for human IGF-I and IGF-II will be described.

---

* For those unfamiliar with the terminology of molecular genetics, a brief explanation is given in the Appendix.

The human IGF-I gene

The IGF-I gene has a discontinuous structure and consists of at least five exons spanning a region of more than 85 kb of chromosomal DNA; Figure 1 (Bell *et al.*, 1985; de Pagter-Holthuizen *et al.*, 1986; Rotwein *et al.*, 1986; de Pagter-Holthuizen *et al.*, 1987). The precise length of the entire gene is still unknown. There is a gap between exons 2 and 3 of more than 59 kb since the cosmids containing the known IGF-I exons do not overlap. Furthermore, because the cDNAs isolated so far are incomplete copies of IGF-I mRNA the 5' end of the gene is not entirely characterized and the promoter is still unidentified.

Analysis of IGF-I cDNA suggests that initiation of translation can occur at three different sites. Three potential initiator methionine (AUG) codons are present at positions –48, –25 and –22 relative to the first amino acid of mature IGF-I. Comparison with the signal peptides from a number of related peptides, e.g. insulin, relaxin and IGF-II favours Met (–25) or Met (–22) as the initiation site.

In vitro translation in a wheat germ translation system of specific IGF-I cDNA transcripts prepared by *in vitro* transcription using Sp6 RNA polymerase, results in two distinct polypeptides which correspond with intiation at Met (–48) and Met (–25) (Figure 2). A third polypeptide band corresponding with initiation at Met (–22) is only faintly visible. This indicates that in a wheat germ system both Met (–48) and Met (–25) can be used as translation initiation codons. However it is not clear whether this system reflects the *in vivo* situation.

Figure 1. Schematic representation of the IGF-I gene, located on the long arm of chromosome 12 and of the two different IGF-I specific cDNAs, IGF-Ia and IGF-Ib. The exons are numbered consecutively 1–5. Asterisks in the 3'–noncoding regions indicate possible polyadenylation signals. Note the difference in scale of genomic DNA compared to the cDNAs.

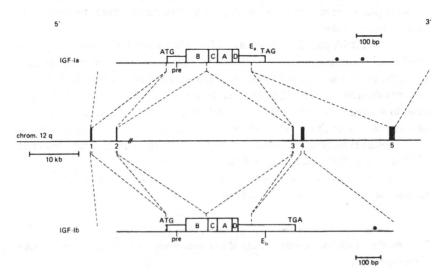

The IGF-I gene can be transcribed into two different mRNA species. Jansen *et al.* (1983) have isolated cDNA from an adult liver library which corresponds to an mRNA consisting of exons 1, 2, 3 and 5 (IGF-Ia). In addition, Rotwein (1986) has isolated a cDNA, IGF-Ib, containing exons 1, 2, 3 and 4, Figure 1, also from an adult liver library. We can conclude that the IGF-I gene encodes both preproIGF-Ia

Figure 2. Polyacrylamide gelelectrophoresis of IGF peptides synthesized in a wheat germ *in vitro* translation system in the presence of ^{35}S–Methionine. Specific IGF-I and IGF-II transcripts were prepared by *in vitro* transcription of cDNA fragments which were cloned in expression plasmids downstream of the Sp6 promoter (Melton *et al.*, 1984).

Lane 1: Translation of IGF-I cDNA transcripts results in polypeptides with MWr 17.0 kD, 15.0 kD and 14.5 kD which correspond with the length of IGF-I precursor peptides initiating at Met (–48), Met (–25) and Met (–22), respectively.

Lane 2: Translation of IGF-II cDNA transcripts results in a single polypeptide band with an IGF-II precursor peptide initiating at Met (–24).

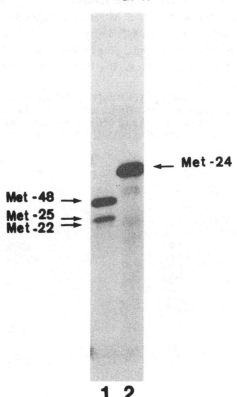

and preproIGF-Ib, which differ only in the amino acid sequence of the carboxyl-terminal E–domain. PreproIGF-Ia contains an E–peptide of 35 amino acids whereas preproIGF-Ib has an E–peptide of 77 amino acids. The corresponding mRNAs must be produced by alternative RNA processing of a single primary transcript. The different combinations of exons code for two distinct E–peptides which may serve tissue specific biological functions.

Another way of creating diversity in IGF-I mRNA species is the use of different polyadenylation signals. This will not influence the peptide itself but may affect RNA stability or translatability.

Expression of IGF-I specific mRNA was studied by Northern blotting, using IGF-I cDNA as a probe (Figure 3). In adult human liver RNA, several IGF-I specific bands can be detected. Two major RNAs of 7.6 kb and 1.1 kb as well as a broad zone of hybridization between 2.5 kb and 5.0 kb are present. In fetal liver a 7.6 kb mRNA is hardly detectable supporting the notion that IGF-I is predominantly involved in postnatal growth. It is not yet clear which of these mRNA transcripts are specific for IGF-Ia or IGF-Ib. As a comparison, adult rat and mouse liver RNA show hybridization patterns similar to the adult human liver (Figure 3). The large size of the predominant IGF-I mRNA species suggests very strongly that the cDNAs isolated so far are not complete. Therefore the entire gene for IGF-I must be larger than the 85 kb of chromosomal DNA which contains the five IGF-I specific exons.

Figure 3. Northern blot of poly(A)⁻ and poly(A)⁺ RNAs isolated from human adult liver (HAL), human fetal liver (HFL), adult rat liver (RL) and adult mouse liver (ML). Each lane contains 10μg of RNA. Hybridization was performed with an IGF-I specific probe containing exons 1, 3, 4 and 5 (Figure 1).

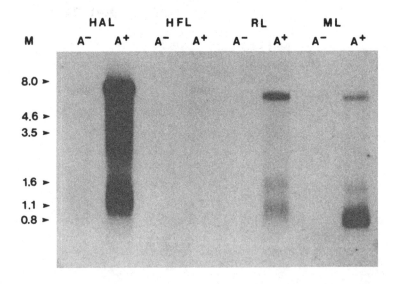

The human IGF-II gene

The IGF-II gene is located on chromosome 11. For the characterization of the IGF-II gene several IGF-II specific cDNAs have been isolated from adult human liver cDNA libraries, none of them however is complete (Jansen *et al.*, 1985). Localization of these adult liver cDNA sequences on cosmid clones showed that the IGF-II gene contains at least six exons. These results imply that IGF-II mRNA from adult liver is transcribed from three 5'-nontranslated exons (exons 1, 2 and 3) and three exons coding for the IGF-II precursor (exons 5, 6 and 7), (Figure 4, de Pagter-Holthuizen *et al.*, 1987).

In our search for new IGF-II cDNAs we also screened a cDNA library from a different source than adult liver, namely the human hepatoma cell line HepG2 (provided by Drs P. Berg and M. McPhaul, Stanford, USA). One IGF-II specific cDNA clone was isolated from the HepG2 cDNA library, which after nucleotide sequence analysis showed an unexpected result (de Pagter-Holthuizen *et al.*, 1987). The cDNA from HepG2 consists of the three coding exons (exons 5, 6 and 7), but is preceded at the 5' end by a nucleotide sequence, diverging from the splice site on, which has not been detected before in other human cDNAs. This sequence is homologous to a rat liver cell line (BRL3A) cDNA sequence determined by Dull *et al.* (1984) and identical to a sequence localized on the human chromosome. The IGF-II cDNA from HepG2 was used as a probe to determine the precise position on the genomic DNA (Figure 4, Dull *et al.*, 1984; de Pagter-Holthuizen *et al.*, 1987).

To establish at which stage in development these two mRNAs are expressed, poly(A)$^+$ RNA was isolated from fetal and adult human liver. Northern blots of poly(A)$^+$ RNA were hybridized with three different ^{32}P-labelled probes. Hybridization with a coding IGF-II probe containing exon 6 to Northern blots with 10 µg poly(A)$^+$ RNA from fetal and adult liver revealed strong expression of a 6.0 kb mRNA in fetal liver, while in adult liver mRNA a weak band of 5.3 kb was detected (Figure 5, lanes 3 and 4). This indicates that the IGF-II gene is predominantly expressed in fetal tissue and confirms the idea that IGF-II is important in fetal growth.

To establish expression of the different 5'-noncoding exons, similar poly(A)$^+$ blots were hybridized with fragments containing exon 1 and exon 4 sequences, respectively. The probe containing exon 1 sequences hybridizes to the 5.3 kb band in adult liver mRNA, but not to the 6.0 kb band in the fetal liver, (Figure 5, lanes 5 and 6). This indicates that exon 1 sequences are only present in adult mRNA and not in fetal mRNA. On the other hand, an exon 4 probe hybridizes only to the 6.0 kb band in fetal mRNA, (Figure 5, lanes 1 and 2). This suggests that exon 4 sequences are only expressed in fetal tissue. It also indicates that two different promoters for the IGF-II gene are present (de Pagter-Holthuizen *et al.*, 1987).

To determine the precise position of initiation of transcription, primer extension experiments were performed with poly(A)$^+$ RNA from adult and fetal liver.

For the fetal mRNA, initiation of transcription takes place at a position 1165 nucleotides upstream of the exon 4 splice donor site. This implies that exon 4 is 1165 bp long and is preceded by a fetal promoter region. Nucleotide sequence analysis of the fetal promoter reveals a TATA–box, a CAAT–box and an Sp1 recognition sequence. These are characteristic elements of most eukaryotic promoters (McKnight & Tjian, 1986).

For the characterization of the adult promoter we have performed similar primar extension experiments with poly(A)+ RNA from adult liver. Initiation of transcription was localized 115 bp upstream of the exon 1 splice donor site. To characterize the adult promoter, the complete nucleotide sequence of the 1.4 kb intergenic region between insulin and IGF-II was determined. This region exhibits some remarkable features. The region upstream of exon 1 does not contain a TATA–box or a CAAT–box. However, an Sp1 recognition site is present and a GC–rich region precedes the site of initiation of transcription. These features have also been described for a number of so-called house-keeping genes which are expressed at low levels in a variety of tissues (Dynan, 1986). Furthermore, the intergenic region contains a number of direct and inverted repeats. Besides several small repeats, an almost perfect 66 bp inverted repeat is present. These repeats may be involved in regulation of expression by interaction with regulatory proteins. Since the nucleotide sequences upstream of exon 1 do not contain typical eukaryotic promoter elements, we have tested whether this region exhibits promoter activity in a biological assay. A

Figure 4. Schematic representation of the human IGF-II gene, located very close to the insulin gene of the short arm of chromosome 11. Two different IGF-II specific cDNAs derived from an adult human liver cDNA library and derived from a human hepatoma (HepG2) cDNA library are indicated. The exons are numbered consecutively 1–7. The structure of the insulin gene was reported by Bell *et al.* (Bell *et al.*, 1980). Note the difference in size of the IGF-II gene compared to the insulin gene.

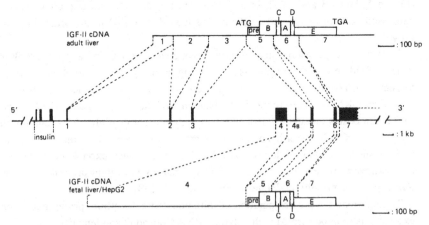

eukaryotic expression plasmid, containing a fragment of the intergenic region inserted in front of a promoter-defective neomycin resistance gene was transfected into a mouse hepatoma cell line. Preliminary results show that the fragment acts as a promoter, confirming the idea that the region upstream of exon 1 is involved in expression of the IGF-II gene in adult liver. Further analysis of the promoter regions for the IGF-II gene is in progress.

Recently, two reports have appeared describing the structure of the rat IGF-II gene (Soares *et al.*, 1986; Frunzio *et al.*, 1986). Comparison of the human IGF-II gene with the rat gene reveals striking homologies and differences between both genes. The rat IGF-II gene contains four exons which are highly homologous to the human exons 4, 5, 6 and 7. However, expression of rat analogues of the human exons 1, 2 and 3 was not detected. The human exons 1, 2 and 3 may therefore be species specific.

In contrast, a new 5'-noncoding exon preceded by a promoter sequence was detected on the rat genome. By comparison of our nucleotide sequences of the human IGF-II gene with the rat sequences of this small 5'-noncoding exon, we can confirm

Figure 5. Northern blot of fetal and adult human liver mRNA.

Poly(A)$^+$ RNA was isolated from human fetal and adult liver. RNAs were size-fractionated on 0.8% agarose gels (10 μg per lane). transferred onto nylon hybridization membranes (Hybond N, Amersham, England) and hybridized to different ^{32}P-labelled probes.

lanes 1 and 2: 852 bp genomic fragment of exon 4.

lanes 3 and 4: 825 bp genomic fragment containing exon 6 sequences.

lanes 5 and 6: 945 bp genomic fragment containing exon 1 sequences.

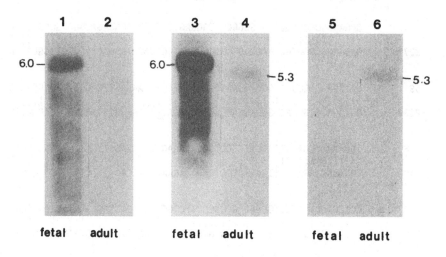

that a homologous sequence is also present on the human genome (Figure 6). We have designated this small human exon, exon 4B (Figure 4). The nucleotide sequence of the region preceding exon 4B contains a TATA-box and a CAAT-box, supporting the notion that the human IGF-II gene actually has three different promoters (Figure 6). We have determined that exon 4B is also expressed in human fetal liver tissue, giving rise to a 4.8 kb mRNA (data not shown).

The results presented indicate that the human IGF-II gene contains three different promoters, which are expressed in a development-specific manner. Up till now, the number of known genes with development-specific expression is still very limited (Leff, Rosenfeld & Evans, 1986). To our knowledge the IGF-II gene is the first human single-copy gene with multiple promoters that are activated in a development-specific way. It is striking that the regulation of expression of the human IGF-II gene seems to differ considerably from the corresponding rat gene. The interpretation of this phenomenon is still unknown.

Tissue-specific expression of the IGF-II gene has been reported previously (Bell *et al.*, 1985). The data presented here indicate that the IGF-II gene can also be expressed in a new development-dependent manner, using different promoters.

It should be noted that the relative activity of the three promoters and the corresponding amounts of mRNA do not imply that equal amounts of the IGF-II precursor protein are synthesized. Differences in the structure of the 5'-untranslated regions can be involved in the efficiency of translation of these mRNAs.

Figure 6. Comparison of the nucleotide sequences of the human IGF-II exon 4B region (upper sequence 1–233) and the corresponding rat sequence (lower sequence 2079–2307; Soares *et al.*, 1986). A potential TATA-box is underlined.The start site of transcription in the rat gene is indicated with an asterisk and the splice donor site in both genes is indicated by ↑.

```
1       GAGAGGGCGGGAGCAAA GGCGCG GGGGAGTGGTCAGCAGGGAGAGGGGTGGGGGGTAG
        ||| |||||||| ||||| || ||| ||||||||||||||| |||| |||| ||||||||
2079    GAGGGGGCGGGTGCAAAGGGGGCGAGGGGAGTGGTCAGCAAGGAGGGGGGGTGGGGGTAG


59      GGTGGAGCCCGGGCTGGGAGGAGTCGGCTCACACAT AAAAGCTGAGGCACTGACCAGCC
        ||||||||| || ||||||||| || |||| |||| ||||| ||||||||||||||
2139    GGTGGAGCCGGGACTGGGAGGAGCCGACTCAGACATAAAAAGCGGAGGCACTGACCAGTT


118     TGCAAACTGGACATTAGCTTCTCCTGTGAAAGAGACTTCCAGCTTCCTCCTCCTCCTCTT
        |||||||||||||| ||||||||||| ||| ||||||| ||| |||| |||
2199    CGCAAACTGGACATTTGCTTCTCCTGTG  AGAACCTTCCAG  CCTTTTCCT GTCT
                *


                                              ↓
178     CCTCCTCCTCCTCCTGCCCCAGCGAGCCTTCTGCT GAGCTGTAG GTAA CA GGCCGT
        || ||||| ||| ||||||||| |||| || | | || || |||| || ||||| |
2252     TCATCCTCTTCCAGCCCCAGCG GCCTCCTTATCCAACTTCAG GTAACCAGGGCCAT
                                                    ↑
```

Summarizing, the analysis of the IGF genes has revealed several levels of regulation of the somatomedin biosynthesis. Modulation is possible at the level of transcription by differential promoter activation and differential splicing. Utilization of different 5'-noncoding exons might affect the efficiency of translation of the corresponding mRNAs and the stability of the mRNAs. Furthermore, alternative polyadenylation sites are used, which can influence the half life of mRNAs. Finally, alternative splicing can cause the formation of different precursor proteins, which may have different biological effects.

In conclusion, we can say that investigation of the IGF-I and IGF-II genes with their remarkabale regulation of expression and alternative processing might lead to the elucidation of various processes involved in development and differentiation.

Acknowledgements

The authors thank F.M.A. van Schaik and R. van der Kammen for excellent technical assistance. They gratefully acknowledge Drs P. Berg and M. McPhaul (Stanford, USA) for the HepG2 cDNA library and Dr G.I. Bell (Emeryville, USA) for communicating unpublished sequence data. This research was supported in part by the foundation for Medical Research (Medigon) with financial aid from the Netherlands Organization for the Advancement of Pure Research (ZWO).

References

Bell, G.I., Pictet, R., Cordell, B., Tischer, E., Goodman, H.M. & Rutter, W.J. (1980). Sequence of the human insulin gene. *Nature*, **284**, 26–32.

Bell, G.I., Merryweather, J.P., Sanchez-Pescador, R., Stempein, M.M., Priestley, L., Scott, J. & Rall, L.B. (1984). Sequence of a cDNA clone encoding human preproinsulin-like growth factor II. *Nature*, **310**, 775–7.

Bell, G.I., Gerhard, D.S., Fong, N.M., Sanchez-Pescador, R. & Rall, L.B. (1985). Isolation of the human insulin-like growth factor genes: insulin-like growth factor II and insulin genes are contiguous. *Proceedings of the National Academy of Sciences, USA*, **82**, 6450–4.

Clemmons, D.R. & Van Wyk, J.J. (1981). Somatomedin C and platelet derived growth factor stimulate human fibroblast replication. *Journal of Cell Physiology*, **106**, 361–7.

Daughaday, W.H., Hall, K., Raben, M.S., Salmon, W.D., Van den Brande, J.L. & Van Wyk, J.J. (1972). Somatomedin: a proposed designation for the "sulfation factor". *Nature*, **235**, 107.

D'Ercole, A.J., Applewhite, G.T. & Underwood, L.E. (1980). Evidence that somatomedin is synthesized by multiple tissues in the fetus. *Developmental Biology*, **75**, 315–28.

Dull, T.J., Gray, A., Hayflick, J.S. & Ullrich, A. (1984). Insulin-like growth factor II precursor gene organization in relation to insulin gene family. *Nature*, **310**, 777–81.

Dynan, W.S. (1986). Promoters for housekeeping genes. *Trends in Genetics*, **2**, 196–7.

Frunzio, R., Chianiotti, L., Brown, A.L., Graham, D.E., Rechler, M.M. & Bruni, C.B. (1986). Structure and expression of the rat insulin-like growth factor II (rIGF-II gene. *Journal of Biological Chemistry*, **262**, 17138–49.

Hintz, R. (1984). Plasma forms of somatomedin and the binding protein phenomenon. In *Clinics in Endocrinology and Metabolism. Tissue Growth Factors*, ed. W.H. Daughaday, Vol. 13, 31–42. W.B. Saunders, Philadelphia.

Höppener, J.W.M., de Pagter-Holthuizen, P., Geurts van Kessel, A.H.M., Jansen, M., Kittur, S.D., Antonarakis, S.E., Lips, C.J.M., & Sussenbach, J.S. (1985). The human gene encoding insulin-like growth factor I is located on chromosome 12. *Human Genetics*, 69, 157–60.

Jansen, M., Van Schaik, F.M.A., Ricker, A.T., Bullock, B., Woods, D.E., Gabbay, K.H., Nussbaum, A.L., Sussenbach, J.S. & Van den Brande, J.L. (1983). Sequence of cDNA encoding human insulin-like growth factor I precursor. *Nature*, 306, 609–11.

Jansen, M., Van Schaik, F.M.A., Van Tol, H., Van den Brande, J.L. & Sussenbach, J.S. (1985). Nucleotide sequence analysis of cDNAs encoding precursors of human insulin-like growth factor II (IGF-II) and an IGF-II variant. *FEBS Letters*, 179, 243–6.

LeBouc, Y., Dreyer, D., Jaeger, F., Binoux, M. & Sondermayer, P. (1986). Complete characterization of the human insulin-like growth factor I nucleotide sequence isolated from a newly constructed adult liver cDNA library. *FEBS Letters*, 196, 108–12.

Leff, S.E., Rosenfeld, M.G. & Evans, R.M.A. (1986). Complex transcriptional units: diversity in gene expression by alternative RNA processing. *Annual Review of Biochemistry*, 55, 1091–1117.

McKnight, S. & Tjian, R. (1986). Transcriptional selectivity of viral genes in mammalian cells. *Cell*, 46, 795–805.

Melton, D.A., Kreig, P.A., Rebagliati, M.R., Maniatis, T., Zinn, K. & Green, M.R. (1984). Efficient *in vitro* synthesis of active RNA and RNA hybridization probes from plasmids containing a bacteriophage SP6 promoter. *Nucleic Acids Research*, 12, 7035–56.

de Pagter-Holthuizen, P., Höppener, J.W.M., Jansen, M., Geurts van Kessel, A.H.M., Van Ommen, G.J.B. & Sussenbach, J.S. (1985). Chromosomal localization and preliminary characterization of the human gene encoding insulin-like growth factor II. *Human Genetics*, 69, 170–3.

de Pagter-Holthuizen, P., Van Schaik, F.M.A., Verduijn, G.M., Van Ommen, G.J.B., Bouma, B.N., Jansen, M. & Sussenbach, J.S. (1986). Organization of the human genes for insulin-like growth factors I and II. *FEBS Letters*, 195, 179–84.

de Pagter-Holthuizen, P., Jansen, M., Van Schaik, F.M.A., Van der Kammen, R,A., Oosterwijk, C., Van den Brande, J.L. & Sussenbach, J.S. (1987). The human insulin-like growth factor II gene contains two development-specific promoters. *FEBS Letters*, 214, 259–64.

Rinderknecht, E. & Humbel, R.E. (1978a). The amino acid sequence of human insulin-like growth factor I and its structural homology with proinsulin. *Journal of Biological Chemistry*, 253, 2769–76.

Rinderknecht, E. & Humbel, R.E. (1978b). Primary structure of human insulin-like growth factor II. *FEBS Letters*, 89, 283–6.

Rotwein, P. (1986). Two insulin-like growth factor I messenger RNAs are expressed in human liver. *Proceedings of the National Academy of Sciences, USA*, 83, 77–81.

Rotwein, P., Pollock, K.M., Didier, D.K. & Krivi, G.G. (1986). Organization and sequence of the human insulin-like growth factor I gene. *Journal of Biological Chemistry*, 261, 4828–32.

Soares, M.B., Turken, A., Ishii, D., Mills, L., Episkopou, V., Cotter, S., Zeitlin, S. & Efstratiadis, A. (1986). Rat insulin-like growth factor II gene. *Journal of Molecular Biology*, 192, 737–52

The General Pathway of Eukaryotic mRNA Formation

A. Structure of an Eukaryotic Gene

Most genes in higher eukaryotes are discontinuous; the coding sequences (exons) are separated by intervening sequences (introns). Other sequences present in eukaryotic genes are the 5' promoter elements (TATA–box and CAAT–box) and at the 3' end the polyadenylation signal.

B. Transcription

RNA polymerase II binds to the promoter site and begins the RNA synthesis. The exact start site for RNA synthesis is the nucleotide to which the cap is added (cap-site). Termination of transcription occurs about 15 to 30 nucleotides downstream from the polyadenylation signal at the polyA-site. Subsequently a polyA tail of about 250 adenylate residues is added to the 3' end of the molecule.

C. Splicing

The final step, called splicing, is the processing of the primary transcript RNA to mature mRNA. The intervening sequences or introns are removed and the remaining pieces, the exons, are joined leading to the finished mRNA product.

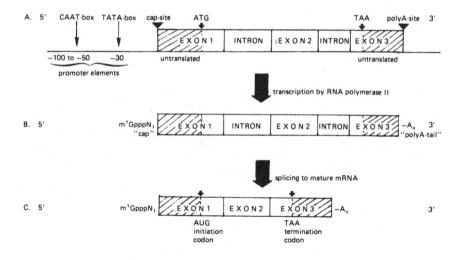

D.J. HILL

Peptide growth factors in fetal development

Introduction

Growth *in utero* consists of a series of integrated biological processes. It is most commonly considered as a time of rapid cellular hyperplasia, yet this is not uniform in all tissues throughout gestation. In the early embryo considerable cell migration and spatial orientation occurs as organogenesis and somite development proceed. As gestation progresses ordered tissue differentiation must occur to prepare the infant for extra-uterine survival, while in the final trimester cell hypertrophy and, in the human, progressive adiposity contribute relatively more to increasing fetal size than does cellular hyperplasia. Peptide growth factors have the potential to drive and coordinate each of these developmental aspects since this class of molecules can variously promote cell hyperplasia, hypertrophy and differentiation; act as chemoattractants to influence cell migration; and, in some circumstances, slow or arrest cell proliferation.

Our conception of peptide growth factors has arisen from studies of postnatal growth, and is ambiguous. The somatomedins/insulin-like growth factors have been traditionally thought to act as liver-derived endocrine hormones which drive longitudinal skeletal growth and whose synthesis is under growth hormone control. Conversely, platelet-derived growth factor has been considered as a peptide whose role is largely limited to wound healing. Such rigid concepts have come increasingly under attack largely due to novel evaluations of growth factor presence and function in the embryo and fetus. These peptides are now considered to be ubiquitously distributed among tissues and to act at or near to their sites of synthesis as paracrine or autocrine messengers. Both peptide synthesis and tissue receptivity vary with ontogeny. This may create a dynamic balance of growth enhancing and inhibiting molecules within the extracellular fluid of each organ and tissue.

Nature of peptide growth factors

Although diverse in structure and specificity peptide growth factors invoke a similar group of cellular responses which include stimulation of glucose and amino acid transport, subsequent RNA and protein synthesis, DNA synthesis and cell replication. Additionally many peptide growth factors will modulate cell differentiation in either a stimulatory or inhibitory manner.

The somatomedins/insulin-like growth factors (SM/IGFs) are single chain polypeptides of approximately 7.5 kD molecular weight with structural similarity to insulin (Blundell et al., 1978). Two SM/IGF peptides have been isolated, SM-C/IGF I and IGF II, of which SM-C/IGF I is the more potent growth-promoting agent in vivo in rats (Schoenle et al., 1985). Both SM/IGF peptides are expressed and released by multiple tissues and most probably act predominantly as paracrine or autocrine agents (Roberts et al., 1987; D'Ercole, Stiles & Underwood, 1984). However, their copious presence in the circulation suggests that an endocrine role may also exist. Tissue expression of the SM-C/IGF I gene, and SM-C IGF I in the circulation, is modulated by nutritional availability and, in postnatal life, circulating growth hormone (GH); the former exerting the primary control (Emler & Schalch, 1987; Roberts et al., 1986). Circulating levels of IGF II are under nutritional control but are less dependent on GH.

The similarities between insulin and SM/IGFs also extend to their cell membrane receptors. Two types of SM/IGF receptor have been identified, a type I receptor, which recognizes SM-C/IGF I with higher affinity than it does IGF II, is structurally related to the insulin receptor and recognizes insulin with low affinity (Jacobs & Cuatrecasas, 1982); and a type 2 receptor which is structurally unrelated to the type I receptor and recognizes IGF II with greater affinity than SM-C/IGF I (Massague & Czech, 1982). The type I receptor mediates those anabolic actions of SM/IGFs so far examained while the type II receptor has not been proven to mediate a biological signal.

The SM/IGFs are complexed with binding proteins both in extracellular fluid and in the circulation. The most widespread form of binding protein has a molecular weight of 35–40 kD and is avidly associated with the membranes of some cells, such as fibroblasts (Clemmons et al., 1986). Here it has the potential to sequestrate SM/IGF peptides. It has also been characterized in serum, amniotic fluid and cerebrospinal fluid. A second form of binding protein is a multi-unit complex of 150 kD restricted to the circulation. The appearance of this binding protein is GH-dependent and it is first detectable in fetal blood after approximately 28 weeks of gestation (D'Ercole, Wilson & Underwood, 1980). Both forms of binding protein extend the biological half-life of SM/IGF peptides.

Epidermal growth factor (EGF) is a single chain polypeptide of molecular weight 6 kD first isolated from the salivary glands of male mice (Cohen, 1962). It has a widespread tissue distribution (Kasselberg et al., 1985) but in the circulation is restricted to platelets, from which it is liberated at sites of tissue damage. While EGF is mitogenic for multiple cell types in vitro it particularly enhances the maturation aof epithelial structures in vivo, promoting skin thickening and keratinization, corneal hyperplasia, premature opening of the eyelids, eruption of teeth, and epithelial cell proliferation throughout the digestive tract (Gospodarowicz, 1981). A structural analogue of EGF is transforming growth factor α (TGFα) with a molecular weight of about 7.5 kD. TGFα reproduces the biological actions of EGF and is recognized with high affinity by the EGF receptor. Some specificity of response does exist since TGFα is a more potent angiogenic factor than EGF (Schreiber, Winkler & Derynck, 1986).

Platelet-derived growth factor (PDGF) is a heterodimeric glycoprotein of molecular weight 31 kD (Deuel & Huang, 1984). PDGF stimulates the proliferation of mesodermal tissues including fibroblasts, smooth muscle cells and chondrocytes. It is stored in the α granules of platelets and released at sites of injury.

Fibroblast growth factor (FGF) has been isolated from bovine pituitary and brain in both an acidic (pI 4.5) and basic (pI 9.6) form, each with molecular weight approximately 16–17 kD (Gospodarowicz *et al.*, 1984). FGF is a highly potent mitogen for vascular endothelial cells, which have also been reported to express the peptide, and for other cell types derived from mesoderm or neuroectoderm including connective tissues, glial cells, adrenal cortex and ovarian granulosa cells. The occurrence of the basic form of FGF in adrenal cortex, corpus luteum, retina, placenta and kidney suggests a widespread distribution.

Transforming growth factor β (TGFβ) is a homodimer of 25 kD molecular weight which acts as a bifunctional growth factor, either stimulating or inhibiting cell proliferation depending on cell type and the method of culture (Roberts *et al.*, 1985). It is abundant in platelets but is also found in placenta, kidney, bone matrix, lymphocytes, and is released by isolated connective tissues during culture. TGFβ has been reported to act as a growth-promoting peptide, particularly for some fibroblast cell lines in soft agar, but to produce a potent growth inhibition of fibroblasts in monolayer and of all epithelial cell types examined regardless of the culture technique. It is released in an inactive form and may be activated by proteolytic cleavage at the target tissue.

Nerve growth factor (NGF) differs from the others in that it has a greater specificity of target tissue, namely peripheral sensory and sympathetic neurones and some neuronal structures of the central nervous system.While not mitogenic, at least following birth, NGF is essential for neurite outgrowth and has chemotactic properties which may aid innervation during early development (Yankner & Shooter, 1982). Following receptor binding NGF is transported through the neurone to the cell body (Thoenen *et al.*, 1985). The predominant site of NGF synthesis is probably innervated tissues such as the iris of the eye, salivary glands, heart and vas deferens. Its presence in the central nervous sytem may be due to synthesis by glial cells (Furukawa *et al.*, 1987).

Considerable synergy exists between different peptide growth factors during the initiation of cell replication. This stems from separate growth factors having precise roles within the cycle of proliferation, PDGF and FGF acting as competence factors for fibroblastic cells enabling cells to enter G_1, while SM/IGFs and EGF are necessary for progression through G_1 to S phase and DNA synthesis (Van Wyk et al., 1981). Growth inhibitory peptides such as TGFβ can block the mitogenic potential of other peptide growth factors (Hill et al., 1986). The extracellular fluid is likely to contain a unique and ontologically variable cocktail of endocrine, paracrine and autocrine hormones and growth factors specific for the proliferative and differentiating requirements of that tissue.

Growth factors in embryology

Peptide growth factors may play a central role in tissue development from early in embryogenesis. A useful tool in these studies has been the availabaility of tumour-derived teratocarcinoma cell lines. These tumours are experimentally obtained from primitive ectoderm of the early embryo and represent undifferentiated, pleuripotential stem cells (Martin, 1980). Teratocarcinoma cell lines will differentiate in vitro, either spontaneously or in response to retinoic acid, to yield derivatives representing all three primitive germ layers. Hence, treatment of the PC 13 teratocarcinoma line with retinoic acid yields cells which resemble extra-embryonic mesoderm of the amnion and yolk sac phenotypically (Rayner & Graham, 1982), while prolonged exposure of the F9 cell line to retinoic acid gives rise to Dif 5 cells with properties of both parietal and visceral endoderm (Nagarajan, Jetten & Anderson, 1983).

The PC13 cell line possesses both type 1 and 2 SM/IGF receptors and will proliferate with a minimal requirement for external growth factors (Heath & Shi, 1986). The F9 cell line contains SM/IGF and insulin receptors, and will proliferate at an enhanced rate in response to either IGF II or insulin (Nagarajan et al., 1982). Upon differentiation with retinoic acid the derivatives of both cell lines become relatively more dependent on exogenous growth factors for their proliferation, and release substantial amounts of IGF II and 40 kD IGF binding protein (Nagarajan et al., 1985; Heath & Shi, 1986). Similarly, EGF and insulin receptor expression was enhanced following the differentiation of embryonal carcinoma cells (Rees, Adamson & Graham, 1979). Therefore, while cells representing primitive ectoderm may already express and utilize peptide growth factors, these hormones may become relatively more important to growth control following gastrulation and the initiation of differentiated cell function.

While teratocarcinoma cell lines provide an excellent experimental tool care must be exercised when extrapolating the findings to normal embryogenesis without the demonstration that fresh embryonic tissues show the same characteristics. The limited evidence available fully supports a peptide growth factor presence and action during early development and tissue pattern formation.

Using cDNA/mRNA hybridization on placental tissue sections Goustin et al. (1985) demonstrated the expression of the proto-oncogene c-sis, which encodes the

B chain of PDGF (Johnson *et al.*, 1984), in human placental trophoblasts from as early as 21 days post-conception. The ability of the trophoblasts to translate PDGF was supported by a release of radioreceptor-assayable peptide from placental explants, and the presence of high affinity PDGF receptors on trophoblast cell lines. The trophoblasts co-expressed the proto-oncogene c-*myc* was also noted in several epithelial regions of the human fetus during the first trimester (Pfeifer-Ohlsson *et al.*, 1985), although the expression of c-*sis* in fetal tissues has, so far, only been noted in preliminary observations (Goustin *et al.*, 1985). These studies reveal three important biological findings with regard to embryonic development. Firstly, peptide growth factors are present and active in the human fetus from the appearance of the first differentiated cell types. Secondly, the co-expression of c-*sis* and c-*myc* by the same cell type in the placenta suggests that PDGF acts as an autocrine agent. Thirdly, the expression of proto-oncogenes forms a normal part of early human development. We have found that proto-oncogene expression extends beyond the first trimester since c-*ras* expression was detectable in multiple tissues at least up to 18 weeks of gestation (Mellersh, Strain & Hill, 1986). A recent report by Slack *et al.* (1987) found that FGF induced the differentiation of primary mesoderm in the early *Xenopus* embryo, and mimicked the actions of an endogenous morphogen.

Both insulin and insulin receptors are present in the fertilized chick embryo from days 3–4 of incubation, prior to the differentiation of the endocrine pancreas (De Pablo *et al.*, 1982; Hendricks, De Pablo & Roth, 1984). In the developing chick embryo insulin stimulated glucose metabolism during the first 30 hours of incubation but had no action during subsequent somite formation (Baroffio *et al.*, 1986). Receptors for SM/IGFs were identified in the chick brain from at least day 2 of incubation (Bassas *et al.*, 1985). This group has also examined the ontogeny of SM/IGF and insulin receptors in the lens epithelium of the chick embryo eye. The two classes of receptor were independently regulated during development, numbers of SM/IGF receptors correlating positively with cellular growth rate while insulin binding capacity was related more to the differentiated state of the cells, decreasing sharply following fibre formation (Bassas *et al.*, 1987). Turning to mammalian species, mRNA for insulin was present in fetal rat yolk sac, and both insulin and pro-insulin were identified by radioimmunoassay of yolk sac tissue extracts (Muglia & Locker, 1984). Studies with cultured embryonic mouse tissues taken on day 9.5 of gestation showed that both IGF II and 40 kD SM/IGF binding protein were released by amnion and extra-embryonic visceral yolk sac (Heath & Shi, 1986). Immunoassayable SM-C/IGF I was released from cultured mouse limb bud mesenchyme removed on day 11 of gestation (D'Ercole, Applewhite & Underwood, 1980). While no direct evidence exists for the human the animal studies strongly suggest that SM/IGFs will be both expressed and be biologically active during early human embryogenesis.

Ontogeny of growth factors and receptors during fetal development

(a) Somatomedins

Animal experiments have shown that SM/IGFs, like insulin and growth hormone, do not readily cross the placenta from mother to fetus (Underwood & D'Ercole, 1984). The lack of umbilical arterial/venous difference in cord blood SM/IGF values at term suggests that the same is true for the human fetus (Franklin et al., 1979). The SM/IGF found within fetal tissues is likely, therefore, to be of endogenous origin, and to be regulated by nutritional and humoral variables from within the fetal compartment.

Human fetal tissues in the first trimester are rich in mRNA transcripts for IGF II at levels comparable to those found in postnatal 'embryonic' tumours of childhood such as Wilm's tumour (Scott et al., 1985). While this study failed to demonstrate the presence of specific SM-C/IGF II mRNA transcripts in early human fetal tissues these have since been reported by Han, D'Ercole and Lund (1987), although in lower abundance than mRNA for IGF II. The ontogeny of mRNA expression in human fetal tissues is unknown, but has been studied in the rat for both SM-C/IGF I and IGF II. We examined IGF II transcripts in thirteen separate tissues from 16 days gestation until adulthood (Brown et al., 1986). All tissues examined, except spleen and pancreas, strongly expressed the IGF II gene during fetal life, but this declined to unmeasurable levels in the adult. However, IGF II transcription was maintained in the adult cerebral cortex. The expression of the IGF II gene declined in a tissue specific manner, occurring prior to birth in lung but not until weaning in the liver. Multiple fetal rat tissues also transcribed the SM-C/IGF I gene which was also more abundant in fetal life than in the adult (Lund et al., 1986). Multiple transcript sizes of mRNA for both SM-C/IGF I and IGF II exist in both the fetus and adult. The occurrence of particular transcript sizes appears to be developmentally regulated in the rat, but the biological relevance of this is not understood.

Both the SM-C/IGF I and IGF II genes are transcribed in the human placenta from first trimester until term (Shen et al., 1986; Mills et al., 1986). The second trimester placenta expressed more mRNA transcripts for IGF II than did placenta from first trimester or term, and mRNA levels were elevated at term in placentae from diabetic mothers compared to those of normal women.

Since it is probable that not all mRNA transcripts for SM/IGFs are translated (Graham et al., 1986) the levels of mRNA cannot be equated directly with the presence of bioactive peptides. SM/IGF peptides are detectable in extracts of human fetal tissues from at least 9 weeks of gestation (D'Ercole et al., 1986) and are widely distributed. When assessed by radioimmunoassay the greatest amounts of SM-C/IGF I were located in lung, intestine, kidney and skin, and least in brain, heart, adrenal and thymus. The cellular distribution of SM/IGF peptides was assessed by immunocytochemistry (Han et al., 1987). They were detected in most organs and tissues but were particularly well represented in hepatocytes, hepatic haemopoietic cells, differentiated skeletal and cardiac muscle fibres, dermal skin, and in epithelial layers of the primitive pulmonary alveoli, the villi of the intestine and the distal convoluted tubules of the kidney. However, antibodies in this study could not fully distinguish between SM-C/IGF I and IGF II. It is of interest that SM/IGF peptides were largely present in differentiated cell types rather than proliferative, stem cell populations. As will be disucssed later, this may be related to the ability of these peptides to enhance tissue differentiation in addition to proliferation.

Recently it has been questioned whether the sites of SM/IGF presence in the human fetus represent the major sites of peptide synthesis. When the cellular distribution of mRNA for SM-C/IGF I and IGF II was assessed by *in situ* hybridization with cDNA probes the mRNA was largely located in fibrous tissues around the organs, stromal tissue within the gut, the dermal skin, and vascular smooth muscle (Han *et al.*, 1987a). Little mRNA was located in the differentiated cell types which contain SM/IGF peptides. This dichotomy may reflect the paracrine nature of SM/IGF action whereby peptides synthesized predominantly by fibrous tissues are sequestered and utilized by adjacent cell types, perhaps by SM/IGF binding protein (Clemmons *et al.*, 1986). Immunocytochemical localization may therefore reflect sites of growth factor action rather than synthesis. However, the ability of differentiated cell types such as human fetal hepatocytes and pancreatic β cells to release SM-C/IGF I during tissue culture suggests that such distinctions are only relative, and that most cells have some capacity to synthesize SM/IGF peptides (Strain *et al.*, 1987; Swenne *et al.*, 1987).

Evidence for a specific, fetal form of human SM-C/IGF I has been reported by Sara *et al.*, (1981) who noted that fetal brain cell membranes could detect SM/IGF quantitatively in infant cord blood that was undetectable by conventional radioimmunoassay or radioreceptor assay using placental membranes. This SM/IGF was isolated from fetal serum and brain as a truncated form of SM-C/IGF I lacking three amino acids from the amino terminal sequence (Sara *et al.*, 1986). Other molecular variations have been reported including some produced by alternative splicing of the precursor mRNA for SM-C/IGF I or IGF II (Rotwein *et al.*, 1986; Zumstein, Luthi & Humbel, 1985). The relevance of all such forms to fetal development requires a demonstration of biological activity.

A developmental change in the presence of SM/IGF peptides occurs in some species during the perinatal period. In both the fetal sheep and rat IGF II is the predominant circulating species of SM/IGF (Daughaday *et al.*, 1982; Gluckman & Butler, 1983) but declines rapidly at parturition. Conversely, circulating levels of SM-C/IGF I are relatively low in fetal life but increase following birth. In the sheep these changes occur very rapidly, within 4–6 days either side of parturition. The ontogeny of circulating SM/IGFs in the human differs in that both peptides are present in the fetus from at least early second trimester (D'Ercole *et al.*, 1986; Ashton *et al.*, 1985) at concentrations 2–3-fold lower than adult values. Circulating levels of IGF II in the fetus were 2–4-fold greater than those of SM-C/IGF I, but the comparative abundance of the former persisted into adult life. Although circulating levels of SM-C/IGF I in the human fetus are relatively low they show a statistically stronger correlation with both skeletal length and fetal body weight than do those for IGF II (Bennet *et al.*, 1983; Gluckman *et al.*, 1983).

In the first trimester numerous human fetal tissues, including lung, liver, brain, heart and kidney possess specific receptors for IGF peptides and for insulin (Potau, Riudor & Ballabriga, 1981; Sara *et al.*, 1983). The observation that fetal brain from early second trimester preferentially bound IGF II, while that obtained between 17 and 25 weeks gestation bound SM-C/IGF I with high affinity, suggests that a developmental change in receptor type may occur. Whether this represents the emergence of the type 1 SM/IGF receptor is not known. Human fetal connective tissues contain both type 1 and 2 SM/IGF receptors from late first trimester and these have been identified in cultures of fetal fibrobasts (Atkinson *et al.*, 1987). Similarly, both receptor species are present on placenta and in liver from early gestation (Daughaday, Mariz & Trevedi, 1981; Bhaumick, Bala & Hollenberg, 1981; Sara *et al.*, 1983; Chernausek *et al.*, 1987). Both SM/IGF and insulin receptors may undergo significant changes in the human fetus near term. Monocytes from cord blood bound five times more insulin, and had a greater SM/IGF binding capacity, than did adult cells (Thorsson & Hintz, 1977; Rosenfeld, Thorsson & Hintz, 1979.

(b) Epidermal growth factor

EGF has been localized by immunocytochemistry in numerous human fetal tissues inthe early second trimester (Kasselberg *et al.*, 1985). These include the pituitary gland, gastric and pyloric glands of the stomach, Brunners' glands of the duodenum, salivary glands, trachea and placenta. While some reports suggest that the analogue of EGF, TGFα is also expressed in the fetal rodent (Twardzik, 1985; Lee *et al.*, 1985) others have argued that TGFα is primarily derived from extra-embryonic sources, such as decidua (Han *et al.*, 1987).

EGF receptors are present on primary ectoderm within outgrowths of whole, cultured blastocysts (Adamson & Meek, 1984). The abundance of trophoblastic cells in the functional placenta may therefore contribute to the high specific binding of EGF to particulate placental membrane preparations, which increases per unit tissue with gestational age (Lei & Guyda, 1984). The EGF receptors are predominantly located on the microvillus plasma membranes of placenta exposed to the maternal circulation, and in the basolateral membranes in close proximity to the fetal circulation (Roa *et al.*, 1984). This is of interest with regard to the reported ability of EGF to induce a release of placental lactogen following infusion into the fetal lamb (Thorburn *et al.*, 1981). Receptors for EGF are also located on human chorion, amnion and decidua (Roa *et al.*, 1984).

Within the mouse fetus from 11 days gestation, EGF binding and receptor number per unit tissue increased until term (Adamson & Meek, 1984). However, the development of EGF receptors is complicated: receptor affinity declined with gestational age for somatic tissues while remaining constant in liver and brain.

(c) Other growth factors

The literature on other peptide growth factors is limited, but does suggest that TGFβ and NGF may contribute to fetal development.

TGFβ bioactivity has been identified in fetal mouse tissues and in conditioned culture medium from isolated fetal rat calvaria and cultured myoblasts from skeletal muscle (Proper, Bjornson & Moses, 1982; Centrella & Canalis, 1985; Hill, Strain & Milner, 1986). In the human fetus TGFβ has been identified in extracts of placenta and in conditioned medium from cultured fibroblasts (Frolik *et al.*, 1983; Lawrence *et al.*, 1984; Elstow *et al.*, 1985).

While it has been speculated that NGF is involved with early neuronal proliferation its presence in fetal life has so far only been confirmed in the placenta and in conditioned medium from chick embryo heart cells (Goldstein, Reynolds & Perez-Polo, 1978; Norrgren, Ebendel & Wikstrom, 1984).

Biological actions of growth factors *in utero*

Assessment of the biological actions of growth factors in the embryo and fetus can be split into two categories. Firstly there is the ability of peptide growth factors, singly or in synergy, to promote cellular hyperplasia. Secondly, there are the wider implications of growth factor action which may include tissue maturation, the acquisition of differentiated function, or programmed cell migration.

(a) Cellular hyperplasia

Insulin, at physiological concentrations, directly stimulates DNA synthesis by teratocarcinoma cell lines and mesenchymal tissues taken from 9.5 day gestation fetal mice (Nagarajan & Anderson, 1982; Heath & Rees,1985). However, a direct mitogenic action of insulin may not persist in fetal life. We found that insulin at physiological or pharmacological concentrations was unable to promote DNA synthesis by monolayers of human fetal fibroblasts obtained in the early second trimester (Atkinson *et al.*, 1987). By contrast, insulin was a potent promotor of amino acid transport by similar cell cultures (Hill *et al.*, 1986). This is consistent with the previous observations of Cheek, Brayton & Scott (1974) and Zetterberg, Engstrom & Dafgard (1984) that insulin predominantly induced cell hypertrophy in the fetus without an increase in hyperplasia.

Unlike insulin, both SM-C/IGF I and IGF II stimulate DNA synthesis and amino acid transport by isolated human fetal connective tissues with half-maximal effective concentrations between 1 and 3 nM (Hill *et al.*, 1986a; Conover, Rosenfeld & Hintz,

1986). SM-C/IGF I and IGF II were biologically equipotent as anabolic agents for human fetal fibroblasts and myoblasts. This is not the case for similar studies with adult human fibroblasts, or for the relative potencies of the two peptides in promoting skeletal growth within the hypophysectomized rat, where SM-C/IGF I is substantially more potent as a growth-promoting agent than is IGF II (Conover et al., 1986b; Schoenle et al., 1985). These different responses of fetal and postnatal tissues may reflect changes in the binding characteristics of the type 1 SM/IGF receptor with age, the type 2 receptor playing no part in the mitogenic signal in either postnatal or fetal fibroblasts (Conover et al., 1986a; Atkinson et al., 1987). SM/IGF peptides have also been shown to stimulate the proliferation of isolated human fetal chondrocytes and DNA synthesis by isolated hepatocytes (Vetter et al., 1986; Strain et al., 1987).

Exogenous EGF promoted DNA synthesis in isolated amnion, limb bud and lung taken from 15-day mouse embryos (Adamson, Deller & Warshaw, 1981), by fetal rat cartilage explants removed late in gestation (Hill et al., 1983) and by human fetal fibroblasts obtained from skin explants in the second trimester (Hill et al., 1986b). Studies with limb bud-derived mesenchymal stem cells from the fetal mouse demonstrated clearly that synergistic interactions occurred between EGF and other peptide growth factors (Kaplowitz, D'Ercole & Underwood, 1982). In high density cultures EGF, SM-C/IGF I and IGF II each induced an increase in cell number, but were more effective in the presence of an uncharacterized fetal mouse liver-derived growth factor. In later fetal development we found that EGF and SM-C/IGF I were synergistic during DNA synthesis by fetal rat cartilage explants (Hill et al., 1983b). Administration of EGF to the fetal lamb in the third trimester caused hypertrophy of the skin and an increase in the weight of the adrenal, thyroid, liver and kidney compared to control animals (Thorburn et al., 1981). However, since EGF does not normally circulate in the plasma the experiment is by necessity unphysiological.

The biological properties of TGFβ differ somewhat from those of the other growth factors since TGFβ is a bifunctional regulator of human fetal connective tissue proliferation, stimulating or inhibiting fibroblast replication in vitro (Hill et al., 1986b). Stimulation was noted when cells were derived from fetuses of less than 50 g body weight, and inhibition was seen when the donor fetus weighed more than 100 g. Both types of response were maximal in the presence of approximately 40 pM of TGFβ. Additionally, when human fetal fibroblasts were exposed to other growth factors such as SM-C/IGF I, EGF or PDGF, the resulting increase in DNA synthesis was either potentiated or inhibited by TGFβ, according to the weight of the donor fetus. The physiological basis of such a bifunctional response is unknown. By contrast, TGFβ is an extremely potent inhibitor of DNA synthesis by isolated human fetal hepatocytes, with a half-maximal concentration of 0.7 pM (Strain, Hill & Milner, 1986). The inhibitory influence is totally reversible upon removal of TGFβ and is independent of fetal size.

Multiple peptide growth factors therefore have the potential to act as paracrine or autocrine regulators in the human fetus. However their actions should not be considered solely in the context of a hyperplastic response, since this may only be part of a larger biological scenario involving tissue differentiation.

(b) Tissue differentiation

Peptide growth factors can alter the pace and direction of tissue differentiation and may synergise or antagonize to yield a dynamic balance of stem cell populations. Three examples of such interactions in the fetus are the maturation of the secondary palate, the differentiation of skeletal muscle, and the ontogeny of glial cell development.

During the formation of the secondary palate bilateral extensions of the maxillary processes fuse medially above the tongue to separate the oral and nasal cavities. Fusion requires that programmed cell death occurs within the medial epithelial lamina allowing the mesodermal elements of the two processes to become confluent (Hassell & Pratt, 1977). This process is complete in the mouse embryo by 17 days gestation.

Exposure of secondary palatal processes to EGF *in vitro* prevents the degeneration of the medial edge palatal epithelium (Grove & Pratt, 1984), due to hypertrophy and keratinization of the medial edge cells. The ability of EGF to alter the course of palatal differentiation is dependent on the presence of palatal mesenchyme, possibly due to both a synergistic, EGF-dependent release of SM-C/IGF I from the mesenchyme (Atkinson, Bala & Hollenberg, 1984) and the synthesis of a specific, EGF-dependent basement membrane between the mesenchymal and epithelial components (Silver, Murray & Pratt, 1984). The growth of the mesenchymal palatal processes is inhibited by the presence of supraphysiological concentrations Grove, 1984). The normal differentiation of the secondary palate may therefore involve a temporal and quantitative interaction between EGF, SM-C/IGF I and glucocorticoids. Other examples of EGF-dependent differentiation include the induction of pulmonary epithelial maturation, surfactant synthesis and lung liquid release in the fetal rabbit or lamb (Catterton *et al.*, 1980; Sundell *et al.*, 1980; Kennedy *et al.*, 1987).

A well-studied model of tissue maturation has been the differentiation of proliferative myoblasts, derived from skeletal muscle, into postmitotic, contractile myotubes. This process is largely completed *in utero* in the human fetus but can still be observed in the first post-natal week in the rat (Dubowitz, 1967). Following the enzymic dispersal and selection of myoblasts differentiation will occur in a time- and density-dependent manner *in vitro*, and can be followed by changes in intracellular levels of myotube-specific enzymes, such as myokinase, or the acquisition of acetylcholine receptors. Additionally myoblast cell lines obtained from the neonatal rat, such as Yaffe's L6 line, will differentiate fully *in vitro* (Ewton & Florini, 1981).

Concentrations of SM-C/IGF I and IGF II in the nM range will promote cell replication in myoblasts from the fetal rat (Hill *et al.*, 1985), human (Hill *et al.*,

1986a) and the L6 cell line (Ewton & Florini, 1980). The SM/IGF peptides also enhance the rate of myoblast differentiation, cells becoming committed to leave the proliferative cycle after two or three cell replications (Ewton & Florini, 1981). The latter action is independent of cell density and the actions of SM/IGFs as mitogens, since differentiation was still observed when cell replication was blocked by the presence of cytosine arabinoside (Ewton & Florini, 1981). The potency of SM-C/IGF I as both a mitogen and a differentiating agent for myoblasts is substantially greater than that of IGF II or insulin, and both actions are mediated by the type 1 SM\IGF receptor (Ewton, Falen & Florini, 1987). Similarly, in differentiated myotube cultures derived from the fetal lamb SM-C/IGF I was more potent than insulin in stimulating protein synthesis thereby contributing to cellular hypertrophy (Harpur, Soar & Buttery, 1987).

Primary and established myoblast cultures also proliferate in response to FGF. However, FGF, in contrast to SM/IGFs, inhibited myoblast differentiation and potentiated the progenitor cell replication (Gospodarowicz et al., 1976; Linkhart, Clegg & Haunscha, 1981). TGFβ, in agreement with its biological actions in many other mesenchymal cell systems, caused a small inhibition in the replication of L6 myoblasts at saturating concentrations (Massague et al., 1986; Florini et al., 1986). Both studies found that TGFβ inhibited myoblast fusion and the biochemical maturation of myotubes. Hence, interactions between separate peptide growth factors may regulate not only myoblast replication but the progression of differentiation.

In view of these findings it is possible to speculate that the burst of muscle differentiation which occurs 7–10 days following birth in the rat (Dubowitz, 1967) may be related both to the rapid decline in tissue bioassayable TGFβ levels which occurs at this time (Hill et al., 1986c), and to the dramatic fall in tissue expression and circulating levels of IGF II, leading to the emergence of SM-C/IGF I, a more potent differentiating agent for myoblasts, as the most abundant species of SM/IGF peptide in the young rat (Brown et al., 1986).

Recently, evidence has been presented to support an interaction between SM/IGF peptides and NGF during the growth and differentiation of fetal neurones and glial cells. Both IGF II and insulin stimulated neurite formation in sensory and sympathetic neurones, as does NGF (Recio-Pinto & Ishii, 1984); while IGF II also enhanced the levels of mRNA for the axonal microfilament component, tubulin in a human neuroblastoma cell line (Mill, Chao & Ishii, 1985). IGF II may be necessary for the maintenance of NGF receptors on the neuroblastoma cells (Recio-Pinto, Lang & Ishii, 1984). Within the central nervous system SM-C/IGF I was found to be a potent stimulator of growth and differentiation for oligodendrocytes within a population of mixed, isolated glioblasts from the newborn rat (McMorris et al., 1986). Since oligodendrocytes maintain the myelin sheaths around neurones of the central nervous sytem a paracrine release of SM/IGFs from neurones or glial cells within the brain may contribute to ordered neural differentiation. The Snell dwarf mouse, which has a congenital deficiency of GH release leading to poor SM/IGF synthesis, suffers from hypomyelination (Noguchi *et al.*, 1982). Similarly, treatment of the newborn rat with GH antiserum to block the hormone resulted in poor myelin formation and a deficiency of glial cells (Pelton *et al.*, 1977).

Control of peptide growth factor expression

The variables controlling the expression of peptide growth factors in embryonic and fetal life are poorly understood. Given an optimal availability of nutritional metabolites and oxygen, growth *in utero*, and the expression of peptide growth factors proceed according to a genetically predetermined schedule to yield a neonate functionally capable of extra-uterine survival. However, intra-uterine endocrine abanormality may be associated with a stunting of fetal growth, as in leprechaunism or the infant with transient neonatal diabetes mellitus, or fetal overgrowth as in Beckwith Wiedeman syndrome. Both extremes of fetal size have been associated with inappropriate circulating levels of SM/IGFs. If these peptides act predominantly as paracrine factors the physiological parameter of study should not be circulating concentrations but the microenvironment of extracellular fluid in each organ or tissue. Such conditions are difficult to sample experimentally and most information has been obtained following the tissue culture of various fetal cell types.

a) Growth hormone and related peptides

The central role of pituitary GH in the maintenance of longitudinal skeletal growth, and the expression of tissues SM-C/IGF I postnatally has naturally led to an assessment of its relevance to growth *in utero*. In the human anencephalic lacking pituitary GH, body weight, after correction for cranial loss, falls within the normal range for gestational age or shows, at most, a small degree of growth retardation (Grumbach & Kaplan, 1983; Honnebier & Swaab, 1973). Similarly, experimental decapitation in the fetal rat and rabbit or hypophysectomy in the fetal lamb does not greatly impede growth rate (Bearn, 1968; Liggins & Kennedy, 1968; Jost, 1979; Hill, Davidson & Milner, 1979; Parkes & Hill, 1985). Assessment of SM-C/IGF I and/or IGF II in the human anencephalic during second trimester, or in the decapitated fetal rabbit and hypophysectomized fetal lamb, demonstrated no reduction in circulating levels compared to normal infants or control animals respectively (Hill *et al.*, 1979; Ashton *et al.*, 1985; Gluckman & Butler, 1985).

The independence of fetal SM/IGF synthesis from pituitary GH was reproduced in studies with isolated cells and tissues. GH did not stimulate the release of SM-C/IGF I from human fetal fibroblasts, myoblasts, cartilage explants or explants of pancreas (Hill, Crace & Milner, 1985; Swenne *et al.*, 1987), fetal rat myoblasts (Hill *et al.*, 1984), or IGF II release from fetal rat fibroblasts or hepatocytes (Adams *et al.*, 1983; Richman *et al.*, 1985). However, hGH did cause the release of SM-C/IGF I, and an increase in DNA synthesis, in isolated human fetal hepatocytes at concentrations of

2–10 nM (Strain *et al.*, 1987). This is within the range of circulating values for the human fetus (Grumbach & Kaplan, 1973). The ability of hGH to enhance DNA synthesis by hepatocytes was blocked in the presence of antibody against SM-C/IGF I, demonstrating the mediation of hormone action by a paracrine release of SM/IGF peptides.

While pituitary GH may be of little relevance to the growth of human fetal connective tissues it may contribute to the control of hepatic development. We recently demonstrated that specific receptor sites for hGH were present on particulate membrane fractions from human fetal liver in late first and early second trimester, but were absent from skeletal muscle membranes (Hill, Strain & Freemark, 1987). However, the possible role of GH-like molecules in fetal development may need re-appraisal with the finding that the placenta also releases a GH molecule which differs structurally from the pituitary peptide, and is not recognized by some pituitary GH antisera (Hennen *et al.*, 1985). The placental GH is most probably a product of the hGH-V gene, a gene normally dormant in the pituitary gland (Frankenne *et al.*, 1987).

The inability of GH to promote either SM/IGF release or cell growth in the majority of fetal cell types examined contrasts with a large body of evidence which suggests that the structurally related peptides, placental lactogens have a stimulatory influence on both parameters. Human placental lactogen (hPL) is a polypeptide of molecular weight 21.5 kD released into both maternal and fetal circulations from the syncytiotrophoblasts. Maternal values rise to 6–10 µg/ml by week 34 of gestation and provide a sensitive indicator of fetal growth and well-being in the second trimester. Until recently any influence of hPL on fetal development was thought to be mediated by a mobilization of maternal glucose and fatty acids for fetal consumption (Spellacy *et al.*, 1971). Animal experimentation has suggested that the high circulating values of PL in the third trimester induce a stimulation of maternal SM-C/IGF I release (Daughaday, Trevedi & Kapadia, 1979) which may account for the high serum values of human SM-C/IGF I found in late gestation (Furlanetto *et al.*, 1977). The physiological significance of such a rise is unknown.

There is now evidence from animal studies which shows that PL peptides promote anabolic and metabolic events in isolated fetal tissues. Ovine PL promoted amino acid transport by isolated fetal and neonatal rat diaphragm muscle, and glycogen synthesis and an inhibition of glycogenolysis by fetal rat liver (Hurley *et al.*, 1980; Freemark & Handwerger, 1983; 1984; 1985). In a homologous sytem oPL was found to be one hundred-fold more potent in promoting glycogen synthesis by isolated ovine fetal hepatocytes than was oGH (Freemark & Handwerger, 1986). These actions are probably mediated by specific oPL receptors present on ovine fetal liver throughout the second and third trimester (Freemark, Comer & Handwerger, 1986). Ovine PL, but not hGH or ovine prolactin, increased the release of IGF II from fetal rat fibroblasts (Adams *et al.*, 1983). However, hPL did not cause the release of IGF II from isolated fetal rat hepatocytes (Richman *et al.*, 1985).

We have found that hPL promoted SM-C/IGF I release accompanied by DNA synthesis, in isolated human fetal fibroblasts, myoblasts, hepatocytes, and pancreas explants between 8 and 9 weeks gestation (Hill *et al.*, 1985a; Strain *et al.*, 1987; Swenne *et al.*, 1987a). The mitogenic actions of hPL on fibroblasts and hepatocytes were partly mediated by a release of SM/IGFs since DNA synthesis was blocked by an anti-SM-C/IGF I antibody (Hill *et al.*, 1986a; Strain *et al.*, 1987). The biological actions of hPL may be mediated by specific receptor sites present on human fetal tissues (Hill *et al.*, 1987). Just how relevant hPL is to normal fetal development is not clear. Several case studies have reported the birth of normal-sized infants to mothers apparently lacking immunoreactive hPL (Nielson, Pederson & Kampman, 1979). This does not necessarily negate an anabolic role for hPL since Frankenne *et al.* (1986) have found evidence for a variant form of hPL in the placenta from such women which was not recognized by some hPL antisera.

Insulin and nutrition

Abnormality of insulin secretion is associated with fetal over- or undergrowth. Hence enhanced fetal somatic development has been described in infants with nesidioblastosis (Heitz, Kloppel & Hacki, 1977) or Beckwith-Wiedeman syndrome (Filippi & McKusick, 1970), which are associated with a hypersecretion of insulin; while the infant with transient neonatal diabetes or pancreatic agenesis is growth-retarded with poor muscle mass and little adipose tissue (Schiff, Colle & Stern, 1972; Hill, 1978).

At a superficial level a relationship appears to exist between circulating concentrations of fetal insulin and tissue release of SM/IGFs. An infant born with transient neonatal diabetes mellitus had low circulating insulin and SM-C/IGF I but a normal IGF II. An immediate clinical improvement was seen following treatment with insulin (Blethen *et al.*, 1981). The biological influence of insulin can be almost completely removed in late gestation by experimental total pancreatectomy, and this was perfected in the fetal lamb by Fowden & Comline (1984). A near absence of insulin for three weeks in the third trimester resulted in a 20% reduction in body weight compared to control lambs, accompanied by a significant decrease in circulating SM-C/IGF I (Gluckman *et al.*, 1987). Circulating levels of IGF II, however, were elevated following pancreatectomy.

The unwary may therefore speculate that insulin is directly involved in the maintenance of SM-C/IGF release. However, the influence of insulin on cellular nutrient uptake and utilization must be taken into account. Although circulating levels of glucose, fructose and lactate were slightly greater in the pancreatectomized fetal lamb than in controls (Fowden & Comline, 1984) a state of cellular malnutrition may have prevailed due to a relative failure of insulin-dependent nutrient transport mechanisms. When the availability of fetal nutrients is directly impaired, as occurs following maternal fasting or ligation of the uterine blood vessels in the rat, circulating levels of SM-C/IGF I decline rapidly, accompanied by a reduced insulin secretion (Hill *et al.*, 1983; Vileisis & D'Ercole, 1986). It is therefore experimentally difficult to separate the influence of nutrition from that of insulin action on the control of SM/IGF synthesis, and fetal body growth in general. In view of the failure of insulin to modulate SM/IGF release by fetal rat myoblasts or IGF II release by liver explants directly, it is likely that it is nutritional availability which exerts a major influence on fetal SM/IGF synthesis, and that insulin acts indirectly by stimulating nutrient utilization.

The permissive nature of insulin during SM/IGF synthesis *in utero* is supported by observations of clinical and experimental fetal hyperinsulinaemia. In the infant of the poorly controlled diabetic mother the high birth weight often found is largely accounted for by excess adiposity, although a limited somatic and visceral overgrowth has been reported (Pederson, Bojsen-Moller & Poulson, 1954). Despite hypersecretion of insulin cord blood levels of SM-C/IGF I and IGF II are unaltered or, at best, slightly above normal for gestational age (Susa *et al.*, 1984; Omori *et al.*, 1985). When the syndrome of fetal overgrowth was reproduced experimentally by implantation of insulin loaded osmotic minipumps in the fetal rhesus monkey, an elevation of fetal SM/IGF secretion was only observed under conditions of gross hyperinsulinaemia (Susa *et al.*, 1984).

Conclusions

While there is now considerable evidence to support a major role for peptide growth factors during growth and differentiation of the embryo and fetus our understanding is far from complete. This stems partly from a lack of suitable technology with which to assess growth factor presence and action. If, as seems likely, peptide growth factors are expressed by multiple tissues and act predominantly as paracrine or autocrine messengers it is of limited value to measure circulating concentrations. While it has been possible to directly examine the expression of some growth factor genes in individual fetal tissues the presence of multiple mRNA transcripts, and doubt over their eventual translation, renders extrapolation to synthesized peptide unsure. The direct quantitation of growth factors in individual tissues, and their histological distribution, is also difficult to interpret since some, such as SM/IGFs, may be sequestered by specific binding proteins on the cell surface or in extracellular matrix. Sites of growth factor synthesis, action, or simply accumulation cannot easily be distinguished. While studies *in vitro* have proved useful it will be exceedingly difficult to reproduce the exact paracrine microenvironment of each particular tissue, coupled with correct nutritional and hormonal influence. In short, our understanding of peptide growth factor presence and action *in utero* has just begun to develop.

References

Adams, S.O., Nissley, S.P., Handwerger, S. & Rechler, M.M. (1983). Development patterns of insulin-like growth factor I and II synthesis and regulation in rat fibroblasts. *Nature*, 302, 150–3.

Adamson, E.D., Deller, M.J. & Warshaw, J.B. (1981). Functional EGF receptors are present on mouse embryo tissues. *Nature*, 291, 656–8.

Adamson, E.D. & Meek, J. (1984). The ontogeny of epidermal growth factor receptors during mouse development. *Developmental Biology*, 103, 62–70.

Ashton, I.K & Vesey, J. (1978). Somatomedin activity in human cord plasma and relationship to birth size, insulin, growth hormone and prolactin. *Early Human Development*, 2, 115–22.

Ashton, I.K., Zapf, J., Einshenk, I. & McKenzie, I..Z. (1985). Insulin like growth factors (IGF) I and II in human foetal plasma and relationship to gestational age and foetal size during mid pregnancy. *Acta Endocrinologica, Copenhagen*, 110, 558–63.

Atkinson, P.R., Bala, R.M. & Hollenberg, M.D. (1984). Somatomedin-like activity from cultured embryo-derived cells: partial characerization and stimulation of production by epidermal growth factor (Urogastrone). *Canadian Journal of Biochemical and Cellular Biology*, 62, 1335–42.

Baroffio, A., Raddatz, E., Markert, M. & Kucera, P. (1986). Transient stimulation of glucose metabolism by insulin in the 1-day chick embryo. *Journal of Cellular Physiology*, 127, 288–92.

Bassas, L., De Pablo, F., Lesniak, M. & Roth, J. (1985). Ontogeny of receptors for insulin-like peptides in chick embryo tissues: early dominance of insulin-like growth factor over insulin receptors in brain. *Endocrinology*, 117, 2321–9.

Bassas, L., Zelenka, P.S., Serrano, J. & De Pablo, F. (1987). Insulin and IGF receptors are developmentally regulated in the chick embryo eye lens. *Experimental Cell Research*, 168, 561–6.

Beam, J.G. (1968). The thymus and the pituitary adrenal axis in ancephaly: a correlative between experimental-fetal endocrinology and human pathological observations. *British Journal of Experimental Pathology*, 49, 136–44.

Bennet, A., Wilson, D.M., Lin, F., Nagashima, R., Rosenfeld, A.G. & Hintz, R.L. (1983). Levels of insulin-like growth factors I and II in human cord blood. *Journal of Clinical Endocrinology and Metabolism*, 57, 609–12.

Bhaumick, B., Bala, R.M. & Hollenberg, M.D. (1981). Somatomedin receptor of human placenta: Solubilization, photolabelling, partial purification and comparison with insulin receptor. *Proceedings of the National Academy of Sciences, USA*, 78, 4279–83.

Blethen, S.L., White, N.H., Santiago, J.V. & Daughaday, W.H. (1981). Plasma somatomedins, endogenous insulin secretion and growth in transient neonatal diabetes mellitus. *Journal of Clinical Endocrinology and Metabolism*, 52, 144–7.

Blundell, T.L., Bedarkar, S., Rinderknecht, E. & Humbel; R.E.. (1978). Insulin-like growth factors: a model for tertiary structure accounting for immunoreactivity and receptor binding. *Proceedings of the National Academy of Sciences, USA*, 75, 180–4.

Brown, A.L., Graham, D.E., Nissley, S.P., Hill, D.J., Strain, A.J. & Rechler, M.M. (1986). Developmental regulation of insulin-like growth factor II mRNA in different rat tissues. *Journal of Biological Chemistry*, 261, 13144–50.

Catterton, W.Z., Escobedo, M.B., Sexson, W.R., Gray, M.E., Sundell, H.W. & Stahlman, M.T. (1979). Effect of epidermal growth factor on lung maturation in fetal rabbits. *Pediatric Research*, 13, 104–8.

Centrella, M. & Canalis, E. (1985). Transforming and non-transforming growth factors are present in medium conditioned by fetal rat calvariae. *Proceedings of the National Academy of Sciences, USA*, **82**, 7335–9.

Cheek, D.B., Brayton, J.B. & Scott, R.E. (1974). Overnutrition, overgrowth and hormones (with special reference to the infant born of the diabetic mother). In *Advances in Experimental Biology and Medicine*, ed. A.F. Roche & F. Falkner, **49**, 47–72. Plenum, London.

Chernausek, S.D., Beach, D.G., Banach, W. & Sperling, M.A. (1987). Characteristics of hepatic receptors for somatomedin-C/insulin-like growth factor I and insulin in the developing human. *Journal of Clinical Endocrinology and Metabolism*, **64**, 737–43.

Clemmons, D.R., Elgin, R.G., Han, V.K.M., Casella, S.J., D'Ercole, A.J. & Van Wyk, J.J. (1986). Cultured fibroblast monolayers secrete a protein that alters the cellular binding of somatomedin-C-insulin-like growth factor I. *Journal of Clinical Investigation*, **77**, 1548–56.

Cohen, S. (1962). Isolation of a submaxillary gland protein accelerating incisor eruption and eyelid opening in the newborn animal. *Journal of Biological Chemistry*, **237**, 1555–62.

Conover, C.A., Misra, P., Hintz, R.L. & Rosenfeld, R.G. (1986a). Effect of an anti-insulin-like growth factor I receptor antibody on insulin-like growth factor II stimulation of DNA synthesis in human fibroblasts. *Biochemical Biophysical Research Communications*, **139**, 501–8.

Conover, C.A., Rosenfeld, R.E. & Hintz, R.L. (1986b). Hormonal control of the replication of human fetal fibroblasts: role of somatomedin-C/insulin-like growth factor I. *Journal of Cellular Physiology*, **128**, 47–54.

Daughaday, W.H., Mariz, I.K. & Trevedi, B. (1981). A preferential binding site for insulin-like growth factor II in human and rat placental membranes. *Journal of Clinical Endocrinology and Metabolism*, **53**, 282–8.

Daughaday, W.H., Parker, K.A., Borowsky, S., Trevedi, B., & Kapadia, M. (1982). Measurement of somatomedin-related peptides in fetal, neonatal and maternal rat serum by insulin-like growth factor (IGF) I radioimmunoassay, IGF II radioreceptor assay (RRA), and multiplication stimulating activity RRA after acid–ethanol extraction. *Endocrinology*, **110**, 575–81.

Daughaday, W.H., Trevedi, B. & Kapadia, M. (1979). The effect of hypophysectomy on rat chorionic somatomammotrophin as measured by prolactin and growth hormone radioreceptor assays: possible significance in maintenance of somatomedin generation. *Endocrinology*, **105**, 210–14.

de Pablo, F., Roth, J., Hernandez, E. & Pruss, R.M. (1982). Insulin is present in chick eggs and early chick embryos. *Endocrinology*, **111**, 1909–16.

D'Ercole, A.J., Applewhite, G.T. & Underwood, L.E. (1980). Evidence that somatomedin is synthesized by mutliple tissues in the fetus. *Developmental Biology*, **75**, 315–28.

D'Ercole, A.J., Hill, D.J., Strain, A.J. & Underwood, L.E. (1986). Tissue and plasma concentrations in the human fetus during the first half of gestation. *Pediatric Research*, B20, 253–5.

D'Ercole, A.J., Stiles, A.D. & Underwood, L.E. (1984). Tissue concentration of somatomedin-C: further evidence for multiple sites of synthesis and paracrine/autocrine mechanism of action. *Proceedings of the National Academy of Sciences USA.*, **81**, 935–9.

D'Ercole, A.J., Wilson, D.F. & Underwood, L.E. (1980). Changes in the circulating form of serum somatomedin-C during fetal life. *Journal of Clinical Endocrinology and Metabolism*, **51**, 674–6.

Deuel, T.F. & Huang, J.S. (1984). Platelet-derived growth factor. Structure, function and roles in normal and transformed cells. *Journal of Clinical Investigation*, 74, 669–76.

Dubowitz, V. (1967). Cross-innovation of fast and slow muscle: histochemical, physiological and biochemical studies. In *Exploratory Concepts in Muscular Dystrophy*, ed. A.T. Milka, pp. 164–82. Excerpta Medica, New York.

Elstow, S.F., Hill, D.J., Strain, A.J., Swenne, I., Crace, C.J. & Milner, R.D.G. (1985). Production and partial purification of TGF–B-like activity from early passage human foetal fibroblasts. *Journal of Endocrinology*, 107, [Suppl.] 100.

Emler, C.A. & Schalch, D.S. (1987). Nutritionally-induced changes in hepatic insulin-like growth factor I (IGF I) gene expression in rats. *Endocrinology*, 120, 832–4.

Ewton, D.Z., Falen, S.L. & Florini, J.R. (1987). The type II insulin-like growth factor (IGF) receptor has low affinity for IGF I analogs: pleiotypic actions of IGFs on myoblasts are apparently mediated by the type I receptor. *Endocrinology*, 120, 115–23.

Ewton, D.Z. & Florini, J.R. (1980). Relative effects of the somatomedins, multiplication-stimulating activity, and growth hormone on myoblasts and myotubes in culture. *Endocrinology*, 106, 577–83.

Ewton, D.Z. & Florini, J.R. (1981). Effects of somatomedins and insulin on myoblast differentiation *in vitro*. *Developmental Biology*, 56, 31–9.

Frankenne, F., Hennen, G., Parks, J.S. & Nielsen, P.V. (1986). A gene deletion in the hGH/hCS gene cluster could be responsible for the placental expression of hGH and/or hCS molecules absent in normal subjects. *Endocrinology*, 118, [Suppl.] 128.

Frankenne, F., Rentier-Delrue, F., Scippo, M-L., Martial, J. & Hennen, G. (1987). Expression of the growth hormone variant gene in human placenta. *Journal of Clinical Endocrinology and Metabolism*, 64, 635–7.

Franklin, R.C., Pepperell, R.J., Rennie, G.C. & Cameron, D.P. (1979). Acid-ethanol-extractable non-suppressible insulin-like activity (NSILA–S) during pregnancy and the puerperium, and in cord serum at term. *Journal of Clinical Endocrinology and Metabolism*, 48, 695–9.

Freemark, M., Comer, M. & Handwerger, S. (1986). Placental lactogen and growth hormone receptors in sheep liver: striking differences in ontogeny and function. *American Journal of Physiology*, 251, E328–33.

Freemark, M. & Handwerger, S. (1984). Ovine placental lactogen stimulates glycogen synthesis in fetal rat hepatocytes. *American Journal of Physiology*, B246, E21–5.

Freemark, M. & Handwerger, S. (1985). Ovine placental lactogen inhibits glucagon-induced glycogenolysis in fetal rat hepatocytes. *Endocrinology*, 116, 1275–80.

Freemark, M. & Handwerger, S. (1986). The glycogenic effects of placental lactogen and growth hormone on ovine fetal liver are mediated through binding to specific fetal ovine placental lactogen receptors. *Endocrinology*, 118, 613–8.

Filippi, G. & McKusick, V.A. (1970). The Beckwith-Wiedemann syndrome. *Medicine*, 49. 279–98.

Florini, J.R., Roberts, A.B., Ewton, D.Z., Falen, S.L., Flanders, K.C. & Sporn, M.B. (1986). Transforming growth factor-B. A very potent inhibitor of myoblast differentiation, identical to the differentiation inhibitor secreted by buffalo rat liver cells. *Journal of Biological Chemistry*, 261, 16509–13.

Fowden, A. & Comline, R.S. (1984). The effects of pancreatectomy on the sheep fetus in utero. *Quarterly Journal of Experimental Physiology*, 69, 319–30.

Frolik, C.A., Dart, L.L. Meyers, C.A., Smiths, D.M. & Sporn, M.B. (1983). Purification and initial characterization of a type-α transforming growth factor from human placenta. *Proceedings of the National Academy of Sciences, USA*, **80**, 3676–80.

Furlanetto, R.W. Underwood, L.E., Van Wyk, J.J. & Handerger, S. (1978). Serum immunoreactive somatomedin-C is elevated in late pregnancy. *Journal of Clinical Endocrinology and Metabolism*, **47**, 695–8.

Furukawa, S., Furukawa, Y., Satoyoshi, E. & Hayashi, K. (1987). Synthesis/secretion of nerve growth factor is associated with cell growth in cultured mouse astroglial cells. *Biochemical Biophysical Research Communications*, **142**, 395–402.

Gluckman, P.D., Barrett-Johnson, J.J., Butler, J.H., Edgar, B. & Gunn, T.R. (1983). Studies of insulin-like growth factor I and II by specific radioligand assays in umbilical cord blood. *Clinical Endocrinology*, **19**, 405–13.

Gluckman, P.D. & Butler, J.H. (1982). parturition related changes in insulin-like growth factors-I and -II in the perinatal lamb. *Journal of Endocrinology*, **99**, 223–32.

Gluckman, P.D. & Butler, J.H. (1985). Circulating insulin-like growth factors-I and -II concentrations are not dependent on pituitary influences in the midgestation fetal sheep. *Journal of Developmental Physiology*, **7**, 405–9.

Gluckman, P.D., Butler, J.H., Comlin, R. & Fowden, A. (1987). The effects of pancreatectomy on the plasma concentrations of insulin-like growth factors 1 and 2 in the sheep fetus. *Journal of Developmental Physiology*, **9**, 79–88.

Goldstein, L.D., Reynolds, C.P. & Perez-Polo, J.R. (1978). Isolation of human nerve growth factor from placental tissue. *Neurochemical Research*, **3**, 175–83.

Gospodarowicz, D. (1981). Epidermal and nerve growth factors in mammalian development. *Annual Reviews in Physiology*, **43**, 251–63.

Gospodarowicz, D., Cheng, J., Lui, G.-M., Baird, A. & Bohlent, P. (1984). Isolation of brain fibroblast growth factor by heparin-sepharose affinity chromatography: identity with pituitary fibroblast growth factor. *Proceedings of the National Academy of Sciences USA*, **81**, 6963–7.

Gospodarowicz, D., Weseman, J., Moran, J.S. & Lindstrom, J. (1976). Effect of fibroblast growth factor on the division and fusion of bovine myoblasts. *Journal of Cell Biology*, **70**, 395–405.

Goustin, A.S. Betsholtz, C., Pfeifer-Ohlsson, S., Persson, H., Rydnert, J., Bywater, M., Holmgren, G., Heldin, C.-H., Westermark, D. & Ohlsson, R. (1985). Co-expression of the *sis* and *myc* proto-oncogenes in developing human placenta suggests autocrine control of trophoblast growth. *Cell*, **41**, 301–12.

Graham, D.E., Rechler, M.M., Brown, A.L., Frunzio, R., Romanus, J.A., Bruni, C.B., Whitfield, H.J., Nissley, S.P., Seelig, S. & Berry, S. (1986). Coordinate developmental regulation of high and low molecular weight mRNAs for rat insulin-like growth factor II. *Proceedings of the National Academy of Sciences USA.*, **83**, 4519–23.

Grove, R.I. & Pratt, R.M. (1984). Influence of epidermal growth factor and cyclic AMP on growth and differentiation of palatal epithelial cells in culture. *Developmental Biology*, **106**, 427–37.

Grumbach, M.M. & Kaplan, S.L. (1973). Ontogenesis of growth hormone, insulin, prolactin and gonadotropin secretion in the human fetus. In *Foetal and Neonatal Physiology*, ed. R.S. Comline, K.W. Cross, E.S. Dawes & P.W. Nathanielsz, pp. 462–87. Cambridge University Press, London.

Han, V.K., D'Ercole, A.J. & Lund, P.K. (1987a). Identification of cellular sites of synthesis of somatomedin/insulin-like growth factors in the human fetus by *in situ* hybridization histochemistry. *Science*, **236**, 193–7.

Han, V.K., Hill, D.J., Strain, A.J., Towle, A.C., Lauder, J.M., Underwood, L.E. & D'Ercole, A.J. (1987b). Identification of somatomedin/insulin-like growth factor immunoreactive cells in the human fetus. *Pediatric Research*, in press.

Han, V.K., Hunter, E.S. Pratt, R.M., Zendegui, J.G. & Lee, D.C. (1987c). Expression of rat transforming growth factor a mRNA during development occurs predominantly in the maternal decidua. *Molecular and Cellular Biology*, in press.

Harper, J.M.M., Soar, J.B. & Buttery, P.J. (1987). Changes in protein metabolism of ovine primary muscle cultures on treatment with growth hormone, insulin, insulin-like growth factor I or epidermal growth factor. *Journal of Endocrinology*, **112**, 87–96.

Hassell, J.R. & Pratt, R.M. (1977). Elevated levels of cAMP alters the effect of epidermal growth factor in vitro on programmed cell death in the secondary palatal epithelium. *Experimental Cell Research*, **106**, 55–62.

Heath, J.K., Bell, S. & Rees, A.R. (1981). Appearance of functional insulin receptors during the differentiation of embryonal carcinoma cells. *Journal of Cell Biology*, **91**, 293–7.

Heath, J.K. & Rees, A.R. (1985). Growth factors in mammalian embryogenesis. In *Growth Factors in Biology and Medicine*, ed. D. Evered, Ciba Fdn. Symp. **16**, 3–22. Longmans, London.

Heath, J.K. & Shi, W.-K. (1986). Developmentally regulated expression of insulin-like growth factors by differentiated murine teratocarcinomas and extra-embryonic mesoderm. *Journal of Embryology and Experimental Morphology*, **95**, 193–212.

Heitz, P.V., Kloppel, G. & Hacki, W.H. (1977). Nesidioblastosis: the pathological basis of persistent hyperinsulinaemic hypoglycaemia in infants. *Diabetes*, **26**, 632–42.

Hendricks, A., De Pablo, F. & Roth, J. (1984). Early developmental and tissue-specific patterns of insulin binding in chick embryo. *Endocrinology*, **115**, 1315–23.

Hennen, G., Franakenne, F., Closset, J., Gomez, F., Pirens, G. & El Khayat, N. (1985). A human placental GH: increasing levels during second half of pregnancy and pituitary GH suppression as revealed by monoclonal antibody radioimmunoassay. *International Journal of Fertility*, **30**, 27–33.

Hill, D.E. (1978). Effect of insulin on fetal growth. *Seminars in Perinatology*, **2**, 319–28.

Hill, D.J., Crace, C.J. & Milner, R.D.G. (1985a). Incorporation of [^{3}H] thymidine by isolated fetal myoblasts and fibroblasts in response to human placental lactogen (HPL): possible mediation of HPL action by release of immunoractive SM-C. *Journal of Cellular Physiology*, **125**, 337–44.

Hill, D.J., Crace, C.J., Nissley, S.P., Morrell, D., Holder, A.T. & Milner, R.D.G. (1985). Fetal rat myoblasts release both rat somatomedin-C (SM-C)/insulin like growth factor I (IGF-I) and multiplication stimulating activity in vitro: partial characterization and biological activity of myoblast-derived SM-C/IGF-I. *Endocrinology*, **117**, 2061–72.

Hill, D.J., Crace, C.J. Strain, A.J. & Milner, R.D.G. (1986a). Regulation of amino acid uptake and deoxyribonucleic acid synthesis in isolated human fetal fibroblasts and myoblasts: effect of human placental lactogen, somatomedin-C, multiplication stimulating activity and insulin. *Journal of Clinical Endocrinology and Metabolism*, **62**, 753–60.

Hill, D.J., Davidson, P. & Milner, R.D.G. (1979). Retention of plasma somatomedin activity in the fetal rabbit following decapitation in utero. *Journal of Endocrinology*, **81**, 93–102.

Hill, D.J., Fekete, M., Milner, R.D.G., De Prins, F. & Van Assche, A. (1983a). Reduced plasma somatomedin activity during experimental growth retardation in the fetal and neonatal rat. In *Insulin-like growth factors/somatomedins. Basic chemistry, biology and clinical importance*, ed. E.M. Spencer, pp. 345–52. de Gruyter, Berlin.

Hill, D.J., Holder, A.T., Seid, J., Preece, M.A., Tomlinson, S. & Milner, R.D.G. (1983b). Increased thymidine incorporation into fetal rat cartilage *in vivo* in the presence of human somatomedin, epidermal growth factor and other growth factors. *Journal of Endocrinology*, **96**, 489–97.

Hill, D.J., Strain, A.J., Elstow, S.F., Swenne, I. & Milner, R.D.G. (1986b). Bifunctional action of transforming growth factor-B on DNA synthesis in early passage human fetal fibroblasts. *Journal of Cellular Physiology*, **128**, 322–8.

Hill, D.J., Strain, A.J. & Freemark, M. (1987). Presence of specific binding sites for human placental lactogen (hPL) and growth hormone (hGH) on human fetal tissue membranes during second trimester. *Journal of Endocrinology*, **112**, [Suppl.] 122.

Hill, D.J., Strain, A.J. & Milner, R.D.G. (1986c). Presence of transforming growth factor-B-like activity in multiple fetal rat tissues. *Cell Biology International Reports*, **10**, 915–22.

Honnebier, W.J. & Swaab, D.F. (1973). The influence of anencephaly upon intrauterine growth of the fetus and placenta and upon gestation length. *British Journal of Obstetrics and Gynaecology*, **80**, 577–8.

Hurley, T.W., Thadani, P., Kuhn, C.M., Schanberg, S.M. & Handwerger, S. (1980). Differential effects of placental lactogen, growth hormone and prolactin on rat liver ornithine decarboxylase in the perinatal period. *Life Science*, **27**, 2269–75.

Jacobs, S. & Cuatrecasas, P. (1982). Insulin receptors and insulin receptor antibodies: structure function relationships. In *Receptors, Antibodies and Disease*, CIBA Foundation Symposium 90, pp. 82–90. Pitman, London.

Johnsson, A., Heldin, C.-H., Wasteson, A., Westermark, B., Deuel, T.F., Huang, J.S., Seeburg, P.H., Gray, A., Ullrich, A., Scrace, G., Stroobant, P. & Waterfield, M.D. (1984). The c-*sis* gene encodes a precursor to the B chain of platelet-derived growth factor. *EMBO Journal*, **3**, 921–8.

Jost, A. (1979). Fetal hormones and fetal growth. *Contributions to Gynaecology and Obstetrics*, **5**, 1–20.

Kaplowitz, P.B., D'Ercole, A.J. & Underwood, L.E. (1982). Stimulation of embryonic mouse limb mesenchymal cell growth by peptide growth factors. *Journal of Cellular Physiology*, **112**, 353–9.

Kasselberg, A.G., Orth, D.N., Gray, M.E. & Stahlman, M.T. (1985). Immunocytochemical localization of human epidermal growth factor/urogastrone in several human tissues. *Journal of Histochemistry and Cytochemistry*, **33**, 315–22.

Kennedy, K.A., Wilton, P., Mellander, M., Rojas, J. & Sundel, H. (1987). Effect of epidermal growth factor on lung liquid secretion in fetal sheep. *Journal of Developmental Physiology*, **8**, 421–8.

Lawrence, D.A., Pircher, R., Krycere-Martinerie, C. & Jullien, P. (1984). Normal embryo fibroblasts release transforming growth factors in a latent form. *Journal of Cellular Physiology*, **121**, 184–8.

Lee, D.C., Rochford, R., Todaro, G.J. & Villarreal, L.P. (1985). Developmental expression of rat transforming growth factor-α mRNA. *Molecular and Cellular Biology*, **5**, 3644–5.

Lei, W.H. & Guyda, H.J. (1984). Characterization of epidermal growth factor receptors in human placental cell cultures. *Journal of Clinical Endocrinology and Metabolism*, **58**, 344–52.

Liggins, G.C. & Kennedy, P.C. (1968). Effects of electrocoagulation of the fetal lamb hypophysis on growth and development. *Journal of Endocrinology*, **40**, 371–81.

Linkhardt, T.A., Clegg, C.H. & Haunscha, S.D. (1981). Myogenic differentiation in permanent clonal mouse myoblast cell lines: regulation by macro-molecular growth factors in the culture medium. *Developmental Biology*, **86**, 19–30.

Lund, P.K., Moets-Staats, B.M., Hynes, M.A., Simmons, J. G., Jansen, M., D'Ercole, A.J. & Van Wyk, J.J. (1986). Somatomedin-C/IGF I and IGF II mRNAs in rat fetal and adult tissues. *Journal of Biological Chemistry*, **261**, 14539–44.

Martin, G.R. (1980). Teratocarcinomas and mammalian embryogenesis. *Science, N.Y.*, **209**, 768–76.

Massague, J. & Czech, M.P. (1982). The subunit structures of two distinct receptors for insulin-like growth factors I and II and their relationship to the insulin receptor. *Journal of Biological Chemistry* , **257**, 5038–45.

Massague, J., Cheifetz, S., Endo, T. & Nadal-Ginard, B. (1986). Type B transforming growth factor is an inhibitor of myogenic differentiation. *Proceedings of the National Academy of Sciences, USA*. **83**, 8206–10.

Mellersh, H., Strain, A.J. & Hill, D.J. (1986). Expression of proto-oncogenes c-H-*ras* and N-*ras* in early second trimester human fetal tissues. *Biochemical Biophysical Research Communications*, **141**, 510–16.

McMorris, F.A., Smith, T.M., De Salvo, S. & Furlanetto, R.W. (1986). Insulin-like growth factor I/somatomedin C: a potent inducer of oligodendrocyte development. *Proceedings of the National Academy of Sciences, USA*, **83**, 822–6.

Mills, J.F., Chao, M.U. & Ishii, D.N. (1985). Insulin, insulin-like growth factor II, and nerve growth factor effects on tubulin mRNA levels and neurite formation. *Proceedings of the National Academy of Sciences, USA*, **82**, 7126–30.

Mills, N.C., D'Ercole, A.J., Underwood, L.E. & Ilan, J. (1986). Synthesis of somatomedin-C/insulin-like growth factor I by human placenta. *Molecular Biology Reports*, **11**, 231–6.

Muglia, L. & Locker, J. (1984). Extrapancreatic insulin gene expression in the fetal rat. *Proceedings of the National Academy of Sciences, USA*, **81**, 3635–9.

Nagarajan, L. & Anderson, W.B. (1982). Insulin promotes the growth of F9 embryonal carcinoma cells apparently by acting through its own receptor. *Biochemical Biophysical Research Communications*, **106**, 974–80.

Nagarajan, L., Anderson, W.B. Nissley, S.P., Rechler, M.M. & Jetten, A.M. (1985). Production of insulin-like growth factor II (MSA) by endoderm-like cells derived from embryonal carcinoma cells: possible mediator of embryonic cell growth. *Journal of Cellular Physiology*, **124**, 199–206.

Nagarajan, L., Jetten, A.M. & Anderson, W.B. (1983). A new differentiated cell line (DIF 5) derived by retinoic acid treatment of F9 teratocarcinoma cells capable of extracellular matrix production and growth in the absence of serum. *Experimental Cell Research*, **147**, 315–27.

Nagarajan, L., Nissley, S.P., Rechler, M.M. & Anderson, W.B. (1982). Multiplication-stimulating activity stimulates the multiplication of F9 embryonal carcinoma cells. *Endocrinology*, **110**, 1231–7.

Nielsen, P.V., Pederson, H. & Kampmann, E.M. (1979). Absence of human placental lactogen in an otherwise uneventful pregnancy. *American Journal of Obstetrics and Gynecology*, **135**, 322–6.

Noguchi, T., Sugisaki, T., Takamatsu, K. & Tsukada, Y. (1982). Factors contributing to the poor myelination in the brain of the Snell dwarf mouse. *Journal of Neurochemistry*, **39**, 1693–9.

Norrgren, G., Ebendel, T. & Wikstrom, H. (1984). Production of nerve growth-stimulating factor(s) from chick embryo heart cells. *Experimental Cell Research*, **152**, 427–35.

Omori, Y., Minei, S., Shimizu, M., Azuma, K.J., Akihisa, R., Hirata, Y., Wakai, K. & Tsushima, T. (1985). Insulin-like growth factor-I and CPR levels in the umbiical cord blood of newborns from diabetic mothers. *Journal of the Tokyo Womens Medical College*, **5**, 971–8.

Parkes, M.J. & Hill, D.J. (1985). Lack of growth hormone-dependent somatomedins or growth retardation in hypophysectomized fetal lambs. *Journal of Endocrinology*, **104**, 193–9.

Pedersen, J., Bojsen-Moller, B. & Poulsen, H. (1954). Blood sugar in newborn infants of diabetic mothers. *Acta Endocrinologica Copenhagen*, **15**, 33–52.

Pelton, E.W., Grindeland, R.E., Young, E. & Bass, N.H. (1977). Effects of immunologically-induced growth hormone deficiency on myelinogenesis in developing rat cerebrum. *Neurology*, **27**, 282–8.

Pfeifer-Ohlsson, S., Rydnert, J., Goustin, A.S., Larsson, E., Betsholtz, C. & Ohlsson, R. (1985). Cell-type specific pattern of *myc* proto-oncogene expression in developing human embryos. *Proceedings of the National Academy of Sciences USA.*, **82**, 5050–4.

Potau, N., Ruidor, E. & Ballabriga, A. (1981). Insulin receptors in human placenta in relation to fetal weight and gestational age. *Pediatric Research*, **15**, 798–803.

Pratt, R.M., Kim, C.S. & Grove, R.I. (1984). Role of glucocorticoids and epidermal growth factor in normal and abnormal palatal development. *Current Topics in Developmental Biology*, **19**, 81–101.

Proper, J.A., Bjornson, C.L. & Moses, H.L. (1982). Mouse embryos contain polypeptide growth factors capable of inducing a reversible neoplastic phenotype in non-transformed cells in culture. *Journal of Cellular Physiology*, **110**, 169–74.

Rayner, M.J. & Graham, C.F. (1982). Clonal analysis of the change in growth phenotype during embryonal carcinoma differentiation. *Journal of Cell Sciences*, **58**, 331–4.

Recio-Pinto, E. & Ishii, D.N. (1984). Effects of insulin, insulin-like growth factor-II and nerve growth factor on neurite outgrowth in cultured human neuroblastoma cells. *Brain Research*, **302**, 323–34.

Rees, A.R., Adamson, E.D. & Graham, C.F. (1979). Epidermal growth factor receptors increase during the differentiation of embryonal carcinoma cells. *Nature, London*, **281**, 309–11.

Richman, R.A., Benedict, M.R., Florini, J.R. & Toly, B.A. (1985). Hormonal regulation of somatomedin secretion by fetal rat hepatocytes in primary culture. *Endocrinology*, **116**, 180–8.

Roa, C.V., Carman, F.R., Ghegini, N. & Schultz, G.S. (1984). Binding sites for epidermal growth factor in human fetal membranes. *Journal of Clinical Endocrinology and Metabolism*, **58**, 1034–42.

Roberts, A.B., Anzano, M.A., Wakefield, L.M., Roche, N.S., Stern, D.F.& Sporn, M.B. (1985). Type B transforming growth factor: a bifunctional regulator of cellular growth. *Proceedings of the National Academy of Sciences, USA*, **82**, 119–23.

Roberts, C.T., Brown, A.L., Graham, D.E., Seelig, S., Berry, S., Gabbay, K.H. & Rechler, M.M. (1986). Growth hormone regulates the abundance of insulin-like growth factor I RNA in adult rat liver. *Journal of Biological Chemistry*, **261**, 10025–8.

Roberts, C.T., Lasky, S.R., Lowe, W.L., Seaman, W.T. & Le Roith, D. (1987). Molecular cloning of rat insulin-like growth factor I complementary deoxyribonucleic acids: differential messenger ribonucleic acid processing and regulation by growth hormone in extra hepatic tissues. *Molecular Endocrinology*, **3**, in press.

Rosenfeld, R., Thorsson, A.V. & Hintz, R.L. (1979). Increased somatomedin receptor sites in newborn circulating mononuclear cells. *Journal of Clinical Endocrinology and Metabolism*, **48**, 456–61.

Rotwein, P., Pollock, R.M., Didier, D.K. & Krivi, G.G. (1986). Organization and sequence of the human insulin-like growth factor I gene. *Journal of Biological Chemistry*, **261**, 4828–32.

Sara, V.R., Carlsson-Skwirut, C., Andersson, C., Hall, E., Sjogren, B., Holmgren, A. & Jornvall, H. (1986). Characterization of somatomedins from human fetal brain: identification of a variant form of insulin-like growth factor I. *Proceedings of the National Academy of Sciences, USA*, **83**, 4904–7.

Sara, V.R., Hall, K., Misaki, M., Fryklund, L., Christensen, L. & Wetteberg, L. (1983). Ontogenesis of somatomedin and insulin receptors in the human fetus. *Journal of Clinical Investigation*, **71**, 1084–94.

Sara, V.R., Hall, K., Rodeck, C.H. & Wetterberg, L. (1981). Human embryonic somatomedin. *Proceedings of the National Academy of Sciences, USA*, **78**, 3175–9.

Schiffe, D., Colle, E. & Stern, L. (1972). Metabolic and growth patterns in transient neonatal diabetes. *New England Journal of Medicine*, **287**, 119–22.

Schoenle, E., Zapf, J., Hauri, C., Steiner, T. & Froesch, E.R. (1985). Comparison of in vivo effects of insulin-like growth factors I and II and of growth hormone in hypophysectomized rats. *Acta Endocrinologica, Copenhagen*, **108**, 167–74.

Schreiber, A.B., Winkler, M.E. & Derynck, R. (1986). Transforming growth factor-α: a more potent angiogenic mediator than epidermal growth factor. *Science*, B232, 1250–3.

Scott, J., Cowell, J., Robertson, M.E., Priestley, L.M., Wadey, R., Hopkins, B., Pritchard, J., Bell, G.I., Rall, L.B., Graham, L.F. & Knott, T.J. (1985). Insulin-like growth factor-II gene expression in Wilm's tumour and embryonic tissues. *Nature*, **317**, 260–2.

Shen, S.-J., Wang, C.-Y., Nelson, K.K., Jansen, M. & Ilan, J. (1986). Expression of insulin-like growth factor II in human placentas from normal and diabetic pregnancies. *Proceedings of the National Academy of Sciences, USA*, **83**, 9179–82.

Silver, M.H., Murray, J.L. & Pratt, R.M. (1984). Epidermal growth factor stimulates type-V collagen synthesis in cultured murine palatal shelves. *Differentiation*, **27**, 205–8.

Slack, J.M.W., Darlington, B.G., Heath, J.K. & Godsave, S.F. (1987). Mesoderm induction in early *Xenopus* embryos by heparin-binding growth factors. *Nature*, **326**, 197–200.

Spellacy, W.N., Buhi, W.C., Schram, J.D., Birk, S.A. & McCraary, S.A. (1971). Control of human chorionic somatomammotrophin levels during pregnancy. *Obstetrics and Gynecology*, **37**, 567–73.

Strain, A.J., Hill, D.J. & Milner, R.D.G. (1986). Divergent action of transforming growth factor B on DNA synthesis in human foetal liver cells. *Cell Biology International Reports*, **10**, 855–60.

Strain, A.J., Hill, D.J., Swenne, I. & Milner, R.D.G. (1987). The regulation of DNA synthesis in human fetal hepatocytes by placental lactogen, growth hormone and insulin-like growth factor I/somatomedin-C. *Journal of Cellular Physiology*, in press.

166   D.J. HILL

Sundell, H.W., Gray, M.E., Serenius, F.S.,, Escobedo, M.B. & Stahlman, M.T. (1980). Effects of epidermal growth factor on lung maturation in fetal lambs. *American Journal of Pathology*, 100, 707–26.

Susa, J.B., Widness, J.A. Hintz, R., Liu, F., Sehgal, P. & Schwartz, R. (1984). Somatomedins and insulin in diabetic pregnancies: effects on fetal macrosomia in the human and rhesus monkey. *Journal of Clinical Endocrinology and Metabolism*, 58, 1099–1105.

Swenne, I., Hill, D.J., Strain, A.J. & Milner, R.D.G. (1987a). Effects of human placental lactogen and growth hormone on the production of insulin and somatomedin-c/insulin-like growth factor I by human fetal pancreas in tissue culture. *Journal of Endocrinology*, 113, 287–303.

Swenne, I., Hill, D.J., Strain, A.J. & Milner, R.D.G. (1987b). Growth hormone regulation of somatomedin-C/insulin-like growth factor I production and DNA replication in fetal rat islets in tissue culture. *Diabetes*, 36, 288–94.

Thoenen, H., Korshing, S., Heumann, R. & Acheson, A. (1985). Nerve growth factor. In *Growth Factors in Biology and Medicine*, ed. D. Evered, CIBA Foundation Symposium, 116, 113–28. Pitman, London.

Thorburn, G.D., Waters, M.J., Young, I.R., Dolling, M., Buntine, D. & Hopkins, P.S. (1981). Epidermal growth factor: a critical ractor in fetal maturation? In *The Fetus and Independent Life* ed. K. Elliott & J. Whelan, CIBA Foundation Symposium, 86, 171–98. Pitman, London.

Thorsson, A.V. & Hintz, R.L. (1977). Insulin receptors in the newborn. *New England Journal of Medicine*, 297, 908–12.

Twardzik, D.R. (1985). Differential expression of transforming growth factor-α during prenatal development of the mouse. *Cancer Research*, 45, 5413–6.

Underwood, L.E. & D'Ercole, A.J. (1984). Insulin and somatomedins/insulin like growth factors in fetal and neonatal development. *Journal of Clinical Endocrinology and Metabolism*, 13, 69–89.

Van Wyk, J.J., Underwood, L.E., D'Ercole, A.J., Clemmons, D.R. Pledger, W.J., Wharton, W.R. & Loef, E.B. (1981). Role of somatomedin in cellular proliferation. In *Biology of Normal Human Growth*, ed. M. Ritzen, A. Aperia, K. Hall, A. Larsson, A. Zetterberg and R. Zellerstrom, pp. 223–9. Raven Press, New York.

Vetter, U., Zapf, J., Heit, W., Helbing, G., Heinze, E., Froesch, E.R. & Teller, W.M. (1986). Human fetal and adult chondrocytes. Effect of insulin-like growth factors I and II, insulin, and growth hormone on clonal growth. *Journal of Clinical Investigation*, 77, 1903–8.

Vileisis, R.A. & D'Ercole, A.J. (1986). Tissue and serum concentrations of somatomedin-C/insulin-like growth factor I in fetal rats made growth retarded by uterine artery ligation. *Pediatric Research*, 20, 126–30.

Yanker, B.A. & Shooter, E.M. (1982). The biology and mechanism of action of nerve growth factor. *Annual Review of Biochemistry*, 51, 845–68.

Zetterberg, A., Engstrom, W. & Dafgard, E. (1984). The relative effects different types of growth factors on DNA replication, mitosis and cellular enlargement. *Cytometry*, 5, 368–75.

Zumstein, P.P., Luthi, C. & Humbel, R.E. (1985). Amino acid sequence of a variant pro-form of insulin-like growth factor II. *Proceedings of the National Academy of Sciences, USA*, 82, 3169–72.

VICKI R. SARA

Insulin-like growth factors in the nervous system: characterization, biosynthesis and biological role

Introduction

The somatomedins or insulin-like growth factors are a family of growth-promoting peptide hormones consisting of insulin-like growth factors I (IGF-I) and II (IGF-II) as well as their variant forms (Hall & Sara, 1983). The first members of this family to be characterized were isolated from human adult plasma by Rinderknecht and Humbel (1978a,b) and termed IGF-I and IGF-II due to their structural homology to insulin. IGF-I and IGF-II are homologous single chain peptides with intrachain disulphide bridges and consist of 70 and 67 amino acids respectively. Several variant forms of these peptides have been identified at both the cDNA and protein level (Jansen et al., 1985; Zumstein, Lüthi & Humbel, 1985).

The somatomedins were first suggested to play a role in neural function prior to their characterization. Studies of hormonal influences on brain development led to the hypothesis that brain growth was regulated by a growth factor whose production could be stimulated by growth hormone (Sara & Lazarus, 1974; Sara et al., 1974). In attempts to characterize this brain growth factor it became obvious that it was a member of the somatomedin family (Sara et al., 1976). It took many years however until this factor could be isolated and its amino acid sequence established (Sara et al., 1986). During this time it also became clear that the factor was not only involved in the regulation of brain growth but also was involved in the maintenance of the mature nervous system (Sara, Hall & Wetterberg, 1981).

In this review, the role of the somatomedins as neuropeptides will be discussed, with emphasis being placed upon their characterization and biosynthesis as well as their biological role in the nervous system.

Characterization

Somatomedin activity has been detected in extracts of neural tissues from several species (Sara et al., 1982a) as well as in cerebrospinal fluid (Beaton, Sagel & Distiller, 1975; Sara et al., 1982b). However it was not until recently that any neural activity has been purified and its chemical structure characterized. The somatomedin species present in the human brain were the first to be isolated and characterized (Sara *et al.*, 1986; Carlsson-Skwirut *et al.*, 1986). Two peptides which differ in isoelectric point were purified from both the adult and fetal human brain by acid gelchromatography, affinity chromatography, FPLC and finally HPLC. Structural analysis identified these peptides as firstly, a variant form of IGF-I with a truncated NH_2-terminal region and secondly, IGF-II. In contrast to the brain variant IGF-I, brain IGF-II was identical to the IGF-II form in serum rather than to any identified variant. These studies established the truncated variant IGF-I as well as IGF-II as the somatomedin species present in the human brain throughout life.

The truncated variant IGF-I displays potent neurotrophic activity (Sara *et al.*, 1986). In terms of the human brain Type 1 receptor, the truncated variant IGF-I is five-and tenfold more potent in crossreaction rather than intact IGF-I and IGF-II respectively. Variant IGF-I is also more potent than IGF-I and IGF-II in stimulating fetal brain cell DNA synthesis *in vitro*. In spite of enhanced biological activity, variant IGF-I displays weak crossreaction with several IGF-I antibodies. Altered functional and immunoreactivity raises the question of additional structural changes are to be found in variant IGF-I. Recently the carboxyterminal amino acid of variant IGF-I was identified as alanine, identical to that of IGF-I. Thus in spite of lack of the entire amino acid sequence this finding makes it unlikely that there are further structural differences. If this proves correct, then the observed functional changes are the result simply of an aminoterminal truncation. The truncation is unlikely to result in any major distortion or destabilization of the structure. Rather, a change in the exposed surface of the molecule could be expected. Consideration of proposed three-dimensional models of IGF-I (Blundell & Humbel, 1980; Dafgård *et al.*, 1985) suggests a possible uncovering within the A domain, especially around A12, as well as the beginning of the B domain. Exposure of specific regions of the peptide must result in increased potency of crossreaction in the brain receptor. It may prove of special functional importance that, as will be discussed later, the brain Type 1 receptor also displays structural differences to that identified in other tissues.

Biosynthesis

A single gene locus for IGF-I has been located on chromosome 12 (q22–q24.1) and for IGF-II, on chromosome 11 (p15) contiguous with the insulin gene (Bell *et al.*, 1985). These genes are expressed in the brain, however identification of variant IGF-I as the final protein product indicates CNS specific processing from the IGF-I gene. This can occur at either the mRNA or precursorer protein level (Figure 1).

Rotwein (Rotwein, 1986; Rotwein *et al.*, 1986) has shown that different IGF-I mRNAs are generated from a single gene transcript. These IGF-I mRNAs encode two different protein precursors of either 195 or 153 amino acids which contain the 7.5 k IGF-I but differ in the sequence and length of their COOH-terminal extensions. As IGF-I mRNA of approximately 1.1 kilobase has been identified in the human brain (Sandberg *et al.*, personal communication). The cDNA sequence of this IGF-I gene transcript has yet to be established. Neural specific mRNA processing has been shown for other peptides such as calcitonin gene-related peptide. However, the variant IGF-I is unlikely to arise from alternative mRNA processing as the NH_2 terminus of IGF-I is not an intro/exon hinge region (Jansen *et al.*, 1983). Rather variant IGF-I most probably arises from CNS specific proteolytic processing of the precursor protein. Neural specific processing of other neuropeptide precursors such as proopiomelanocortin has been established. Neuropeptides frequently undergo posttranslational modifications, usually at their NH_2 or COOH termini (Loh & Gainer, 1983). Such modifications can provide a regulatory mechanism for neuropeptide potency. It is therefore hypothesized that the NH_2-terminal truncation of IGF-I in the human brain is a posttranslational modification which represents a neurospecific regulatory mechanism to enhance neuroactivity of the peptide.

Although IGF-II has been isolated from both the fetal and adult human brain (Sara *et al.*, 1986; Carlsson-Skwirut *et al.*, 1986), expression of the IGF-II gene has only

Figure 1. It is hypothesized that identification of the truncated variant IGF-I as the final protein product in the brain implies neural specific processing from the IGF-I gene. Whilst this may occur at either the mRNA or precursor protein level, it is suggested that the NH_2-terminal truncation of IGF-I is a posttranslational modification which represents a regulatory mechanism to enhance neuroactivity.

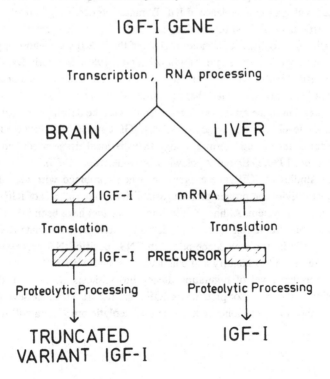

been detected in the adult human brain (Scott *et al.*, 1985). Since this finding also contrasts with several studies identifying abundant IGF-II transcripts in the rat fetal brain, this initial result remains to be confirmed using mRNA preparations with established integrity. In other human fetal tissues, IGF-II transcripts are more abundant than in the adult (Scott *et al.*, 1985).

Expression of both IGF genes appears tissue and developmental specific. Several mRNA species have been identified and their relative abundance varies during development in different tissues. The majority of these studies have been performed in the rat where both IGF-I and IGF-II mRNAs are present in the fetal brain whereas only IGF-II mRNA has been consistently detected in the adult brain (Table 1). Soares, Ishii & Efstratiadis (1985) first reported tissue specific IGF-II mRNA transcripts and demonstrated the predominance of a 1.75/1.6 kb IGF-II mRNA in neonatal rat brain compared to several other tissues such as liver or muscle where the major transcript was 3.4 kb. Although Lund *et al.* (1986) reported IGF-II mRNA in both fetal and adult brain, the predominant species were 4.7 and 3.9 kb similar to other tissues. This discrepancy may be due to the use of different probes; however, caution must be taken with multiple mRNA species until their functional activity can be established. For example, Graham *et al* (1986) have shown that whilst a 4 kilobase IGF-II mRNA predominates in rat liver, it does not direct the translation of pre-pro-r IGF-II in vitro. IGF-II gene expression in the rat brain appears to be developmentally regulated with greater abundance of IGF-II transcripts occurring in fetal compared to adult (Brown *et al.*, 1986; Lund *et al.*, 1986). This may relate to the recent observation of two distinct promotor regions of the IGF-II gene whose appearance is development specific (de Pagter-Holthuizen *et al.*, 1987). Similarly IGF-I mRNA is only consistently observed in fetal rather than adult rat brain. The major brain IGF-I transcript is 1.7 kb similar to other rat tissues (Lund *et al.*, 1986). Thus studies of gene expression in the rat are so far consistent with the development pattern earlier based upon levels of the final gene products, IGF-I and IGF-II, with the latter being the predominant rat fetal form although there is local tissue production of IGF-I (Moses *et al.*, 1980; D'Ercole, Applewhite & Underwood, 1980).

These studies of IGF gene expression, when combined with identification of functional activity and protein products establish the biosynthesis of IGFs within the central nervous system. Although only single gene loci have been identified for the IGFs, variation in the final protein product may arise from alternate RNA splicing pathways. Whilst some evidence points to CNS specific RNA processing, further studies are needed to identify and characterize specific brain IGF transcripts. The second possible level of regulation during biosynthesis arises from differential processing of the precursor proteins for IGF-I and IGF-II. There are several different cleavage sites in these prohormones where proteolytic processing will result in the

Table 1. *Summary of studies reporting IGF mRNA in rat brain. The size (kilobase) is given*

Fetal/Neonatal			Adult	Reference
IGF-I	7.5		ND	Lund *et al.*, 1986
	4.7			
	1.7 *			
	1.2			
IGF-II	3.4		–	Soares *et al.*, 1985
	1.75/1.6 *			
IGF-II	6	Cortex	4*	Brown *et al.*, 1986
	4 *	Hypothalamus	4*	
	3.8	Brain stem	ND	
	2.2			
	1.7			
	1.2			
IGF-II	4.7 *		4.7	Lund *et al.*, 1986
	3.9 *		3.9*	
	2.2			
	1.75			
	1.2			

*major hybridizing band, ND – not detectable.

generation of different peptides. Such post-translational modification is the most likely mechanism for the production of the truncated variant IGF-I in the brain.

Studies of IGF biosynthesis in metabolically labeled cells suggest that the signal peptide is cleaved from the precursor by a microsomal peptidase and that processing of the prohormone occurs at the time of secretion from the cell (Yang *et al.*, 1985). The mechanisms which regulate IGF biosynthesis and secretion in the nervous system are as yet unknown. In contrast to other tissues however, GH does not appear to influence CNS production (D'Ercole, Stiles & Underwood, 1984). T3 has been reported to enhance IGF production mediated by fetal mouse hypothalamic cells but this may be a secondary effect of maturation of the cells (Binoux *et al.*, 1985), alternatively substrates may play a primary role. It has been observed that electrical stimulation of brachial and sciatic nerves leads to a rapid release of somatomedins from cat extirpated limbs, suggesting neural release after depolarization (Sara *et al*, 1982a). More recent studies indicate cell body synthesis and rapid axoplasmic transport of IGF-I in rat sciatic nerve (Hansson, Rozell & Skottner, 1987).

Localization of IGF biosynthesis remains to be established. In earlier studies in the cat, IGF activity was extracted from all CNS areas examined including spinal cord, with the highest content occurring in the cortex and hypothalamus (Sara *et al.*,

1982a). IGF activity was also isolated from sympathetic ganglia, vagus and sciatic nerves. IGF-I has been identified in nerve and Schwann cells in the rat autonomic and peripheral nervous system by immuno-histochemical studies (Andersson *et al.*, 1986). Immunoreactive IGF-I was mostly demonstrated in anterior horn motor neurones, spinal and autonomic ganlion neurons. IGF-II mRNA has been observed in all the rat brain areas examined, i.e. cortex, hypothalamus and brain stem (Brown *et al.*, 1986), however whether these transcripts are translated to functionally active peptides remains to be established. In man, immunoreactive IGF-II is distributed throughout the brain (Haselbacher *et al.*, 1985). The cellular type producing IGFs is also unknown although based on conditioned media studies a glial cell origin for IGF synthesis may be likely.

After their release from the cell, the IGFs are found coupled to carrier proteins. The origin of the binding protein is unknown. Sara & Hall (1984) have suggested that this may be a transmembranal peptidase which proteolytically processes the prohormone to remain bound to the liberated IGF-I or IGF-II. In the cerebrospinal fluid or in glial and neuronal conditioned media IGF appears as a higher molecular weight complex, consisting of IGFs bound to carrier proteins. IGFs can be separated from their carrier proteins by acid gel chromatography (Figure 2). IGF activity was first detected in CSF by Benton *et al.* (1975). Later studies of CSF demonstrated the presence of two forms of immunoreactive IGF-II (9/7 K) (Haselbacher & Humbel, 1982) as well as a binding protein (approx. 40 K) with selective affinity for IGF-II (Binoux, Lassarre & Gourmelen, 1986). An IGF-II specific binding protein has not been observed in glial or neural conditioned medium nor in human brain cytosol.

Receptors

IGF receptors are classified as Type I or Type II receptors (Rechler & Nissley, 1986). The Type I receptor which preferentially recognizes IGF-I consists of two 130 K extracellular α-units and two 98 K transmembranal β-units joined by disulphide bridges. The Type II receptor preferentially recognizes IGF-II and is a 250 K single chain polypeptide. At high concentrations, insulin crossreacts in the Type I receptor which is structurally related to the insulin receptor. Only the Type I receptor is down-regulated by the ligand. In contrast, the Type II receptor is un-regulated by insulin due to redistribution of receptors. The cDNA sequence of the human Type I receptor has been determined and the IGF-I receptor gene mapped to chromosome 15 (q25–26) (Ullrich *et al.*, 1986).

Both Type I and Type II receptors have been identified in the brain of man and rat (Table 2). Both receptors are widely distributed throughout the human brain (Sara *et al.*, 1982b). The Type I receptor appears to predominate in all brain regions where the greatest concentration is found in the cerebral cortex. The ontogenesis of these receptors in the human brain has been followed (Sara *et al.*, 1983). Type I receptors predominate in the human brain where their concentration is greatest during the intrauterine brain growth spurt. The Type I receptor in the brain shows structural differences to that in other tissues, having a smaller α-subunit of approximately 120 K whereas the Type II receptor in the brain is similar to that found in other tissues having a molecular weight of 250 K (Gammeltoft *et al.*, 1985; Roth *et al.*, 1986). The structural difference in the brain Type I receptor is mainly due to differences of glycosylation, since removal of carbohydrate moeities results in identical size of the protein backbone of the α-subunits (Heidenreich *et al.* 1986). This difference in the carbohydrate domains of the hormone binding site containing α-subunit may have special functional significance by affecting the tertiary structure and membrane orientation of the receptors.

Biological action

The IGFs have been demonstrated to have a direct biological action on neural and glial cells in vitro. These studies have used either primary suspensions of rat fetal or neonatal brain cells or human neuroblastoma cells. In these cell culture systems, variant IGF-I and IGF-II have been shown to stimulate DNA synthesis and cell proliferation (Enberg, Tham & Sara, 1985), neurite formation and outgrowth

Figure 2. Separation of cerebrospinal fluid by acid gelchromatography. Brain radioreceptorassay activity (RRA-IGF-I) elutes as three peake, representing binding protein (>30K) and IGF (APPROX. 10K AND 7K)

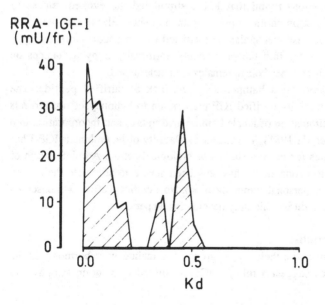

Table 2. *Summary of studies showing brain IGF receptors*

Species	Type	Reference
Human		
Fetal	I,II	Sara *et al.*, 1983
Adult	I,II	Sara *et al.*, 1982
Adult	I,II	Gammeltoft *et al.*, 1985
Adult	I	Roth *et al.*, 1986
Adult	I	Heidenreich *et al.*, 1986
Rat		
Fetal	I	Van Schravendijk *et al.*, 1986
Adult	I,II	Goodyer *et al.*, 1984
Adult	I,II	Gammeltoft *et al.*, 1985
Adult	I	Lowe & Le Roith, 1986

(Recio-Pinto, Lang & Ishii, 1984), as well as neuronal and glial enzyme activity (Lenoir & Honegger, 1983). Recently, McMorris *et al.* (1986) demonstrated that IGF-I was a potent inducer of oligodendrocyte development, thus implicating IGF-I in the myelination of the CNS. Apart from an action on neural and glial growth and differentiation an effect on adult brain function has recently been demonstrated. In studies of acetylcholine synthesis and release in adult rat brain slices, Nilsson *et al.* (personal communication) found that IGF-I stimulated the evoked release of acetylcholine. No effect on choline uptake or unstimulated release was observed. Whether this IGF-I release mechanism is mediated by ion conductance changes, stimulation of Ca^{2+} entry into the pre-synaptic terminals, or by influences on acetylcholine synthesis and metabolism remains to be determined.

In vivo studies have been hampered by the lack of purified peptide. The administration of a partially purified IGF preparation to hypophysectomized rats produced a potent stimulation of labeled amino acid uptake and incorporation into brain protein (Sara *et al.*, 1981). The recent availability of recombinant IGF-I has provided opportunities for in vivo studies of biological action. Administration of recombinant IGF-I to neonatal rats has now been shown to stimulate their brain growth (Philipps *et al.*, personal communication). No significant effect was observed with GH administration during this early developmental period.

Clinical studies

In accordance with their role as growth and maintenance hormones in the CNS, clinical studies suggest a role for IGFs in disorders of brain atrophy and degeneration as well as those arising from disturbances in brain development (Sara, 1986). For example, CSF levels are reduced in chronic alcoholic patients and relate to brain atrophy as determined by computed tomography (Tham *et al.*, 1986). This may not be restricted to the CNS as Hansson *et al.* (1986) have suggested that IGF-I is involved in the regeneration of rat sciatic nerves. The IGFs have also been implicated in mental retardation resulting from disorders of brain development. In Down syndrome for example, a delay in fetal IGF synthesis during the critical intrauterine period of neural formation and migration has been suggested (Sara *et al.*, 1983). In these patients after birth serum IGF-I levels fail to rise during childhood, resulting in growth retardation. IGFs have been identified in brain tumours, especially glioblastoma where, additionally, an enhanced expression of Type II receptors is observed (Sara *et al.*, 1986). It has been suggested that IGF levels in CSF or cystfluid may provide a useful diagnostic tool to assess glioblastoma activity (Prisell *et al.*, 1987).

Conclusion
The IGFs within the nervous system have been characterized as truncated variant IGF-I as well as IGF-II. These neuropeptides are produced within the nervous system throughout life. Their biosynthetic pathways, especially that for variant IGF-I, presumably involve neural specific RNA and prohormone processing. Further studies are essential to define each step in their biosynthesis as well as in cellular localization and regulation. Both Type I and II receptors have been identified and shown to be widely distributed throughout the nervous system. The structural difference observed in the Type I receptor also indicates neural specific receptor biosynthesis, and may have special functional importance, especially in view of the enhanced potency of variant IGF-I. Within the nervous system, the IGFs appear to have a paracrine anabolic action, acting as growth and maintenance neuropeptides. This implies an important role not only in the development of the nervous system but also in disorders of neural degeneration and brain atrophy. Regulation of neural metabolism may additionally modulate cellular sensitivity and responsiveness.

Acknowledgements
These studies have been supported by the Swedish Medical Council, Sävstaholmsföreningen, and Loo and Hans Osterman Fund.

References
Andersson, I., Billig, H., Fryklund, L., Hansson H.-A., Isaksson, O., Isgaard, J., Nilsson, A., Rozlell, B., Skottner, A. & Stemme, S. (1986). Localization of IGF-I in adult rats. Immunohistochemical studies. *Acta Physiologica Scandinavica*, **126**, 311–2.
Beaton, G.R., Sagel, J. & Distiller, L.A. (1975). Somatomedin activity in

cerebrospinal fluid. *Journal of Clinical Endocrinology and Metabolism*, **40**, 736–7.

Bell, G.I., Gerhard, D.S., Fong, N.M., Sanchez-Pescardor, R. & Rall, L.B. (1985). Isolation of the human insulin-like growth factor genes: insulin-like growth factor II and insulin genes are contiguous. *Proceedings of the National Academy of Sciences, USA*, **82**, 6450–4.

Binoux, M., Faive-Bauman, A., Lassarre, C., Barret, A. & Tixier-Vidal, A. (1985). Triiodothyronine stimulates the production of insulin-like growth factor (IGF) by fetal hypothalamus cells cultured in serum-free medium. *Developmental Brain Research*, **21**, 319–21.

Binoux, M., Lassarre, C. & Gourmelon, M. (1986). Specific assay for insulin-like growth factor (IGF) II using the IGF binding proteins extracted from human cerebrospinal fluid. *Journal of Clinical Endocrinology and Metabolism*, **63**, 1151–5.

Blundell, T.L. & Humbel, R.E. (1980). Hormone families: pancreatic hormones and homologous growth factors. *Nature*, **287**, 781–7.

Brown, A.L., Graham, D.E., Nissley, S.P., Hill, D.J., Strain, A.J. & Rechler, M.M. (1986). Developmental regulation of insulin-like growth factor II mRNA in different rat tissues. *The Journal of Biological Chemistry*, **261**, 13144–50.

Carlsson-Skwirut, C., Jörnvall, H., Holmgren, A., Andersson, C., Bergman, T., Lundquist, G., Sjögren, B. & Sara, V.R. (1986). Isolation and characterization of variant IGF-I as well as IGF-II from adult human brain. *FEBS Letters*, **201**, 46–50.

Dafgård, E., Bajaj, M., Honegger, A.M., Pitts, J., Wood, S. & Blundell, T. (1985). The conformation of insulin-like growth factors: relationships with insulins. *Journal of Cell Science*, **3**, 53–64.

D'Ercole, A.J., Applewhite, G.T. & Underood, L.E. (1980). Evidence that somatomedin is synthesized by multiple tissues in the fetus. *Developmental Biology*, **75**, 315–28.

D'Ercoll, A.J., Stiles, A.D. & Underwood, L.E. (1984). Tissue concentrations of somatomedin C: further evidence for multiple sites of synthesis and paracrine or autocrine mechanisms of action. *Proceedings of the National Academy of Sciences, USA*, **81**, 935–9.

Enberg, G., Tham, A. & Sara, V.R. (1985). The influence of purified somatomedins and insulin on fetal brain DNA synthesis in vitro. *Acta Physiologica Scandinavica*, **125**, 305–8.

Gammeltoft, S., Haselbacher, G.K., Humbel, R.E., Fehlmann, M. & Van Obberghen, E. (1985). Two types of receptor for insulin-like growth factors in mammalian brain. *The EMBO Journal*, **4**, 3407–12.

Goodyer, C.G., de Stéphano, L., Lai, W.H., Guyda, H.J. & Posner, B.I. (1984). Characterization of insulin-like growth factor receptors in rat anterior pituitary, hypothalamus, and brain. *Endocrinology*, **114**, 1187–95.

Graham, D.E., Rechler, M.M., Brown, A.L., Frunzio, R., Romanus, J.A., Bruni, C.B., Whitfield, H.J., Nissley, S., Seelig, S. & Berry, S. (1986). Coordinate developmental regulation of high and low molecular weight mRNAs for rat insulin-like growth factor II. *Proceedings of the National Academy of Sciences, USA*, **83**, 4519–23.

Hall, K. & Sara, V.R. (1983). Growth and somatomedins. *Vitamins and Hormones*, **40**, 175–233.

Hansson, H.A., Dahlin, L.B., Danielsen, N., Fryklund, L., Nachemson, A.K., Polleryd, P., Rozell, B., Skottner, A., Stemme, S. & Lundborg, G. (1986). Evidence indicating trophic importance of IGF-I in regenerating peripheral nerves. *Acta Physiologica Scandinavica*, **126**, 609–14.

Hansson, H.A., Rozell, B. & Skottner, A. (1987). Rapid axoplasmic transport of insulin-like growth factor I in the sciatic nerve of adult rats. *Cell and Tissue Research*, **247**, 241–7.

Haselbacher, G. & Humbel, R. (1982). Evidence for two species of insulin-like growth factor II (IGF II and "big" IGF II) in human spinal fluid. *Endocrinology*, **110**, 1822–4.

Hasselbacher, G.K., Schwab, M.E., Pasi, A. & Humbel, R.E. (1985). Insulin-like growth factor II (IGF II) in human brain: regional distribution of IGF II and of higher molecular mass forms. *Proceedings of the National Academy of Sciences, USA*, **82**, 2153–7.

Heidenreich, K.A., Freidenberg, G.R., Figlewicz, D.P. & Gilmore, P.R. (1986). Evidence for a subtype of insulin-like growth factor I receptor in brain. *Regulatory Peptides*, **15**, 301–10.

Jansen, M., van Schaik, F.M.A., Ricker, A.T., Bullock, B., Woods, D.E., Gabbay, K.H., Nussbaum, A.L., Sussenbach, J.S. & Van den Brande, J.L. (1983). Sequence of cDNA encoding human insulin-like growth factor I precursor. *Nature*, **306**, 609–11.

Jansen, M., van Schaik, F.M.A., van Tol, H., Van den Brande, J.L. & Sussenbach, J.S. (1985). Nucleotide sequences of cDNAs encoding precursors of human insulin-like growth factor II (IGF-II) and IGF-II variant. *FEBS Letters*, **179**, 243–6.

Lenoir, D. & Honegger, P. (1983). Insulin-like growth factor I (IGF I) stimulates DNA synthesis in fetal rat brain cell cultures. *Developmental Brain Research*, **7**, 2055–213.

Loh, Y.P. & Gainer, H. (1983). Biosynthesis and Processing of Neuropeptides. In *Brain Peptides*, ed. D.T. Krieger, M.J. Brownstein & J.B. Martin, p.79. John Wiley & Sons, Inc., New York.

Lund, P.K., Moats-Staats, B.M., Hynes, M.A., Simmons, J.G., Jansen, M., D'Ercole, A.J. & Van Wyk, J.J. (1986). Somatomedin-C/insulin-like growth factor-I and insulin-like growth factor-II mRNAs in rat fetal and adult tissues. *The Journal of Biological Chemistry*, **261**, 14539–44.

McMorris, F.A., Smith, T.M., DeSalvo, S. & Furlanetto, R.W. (1986). Insulin-like growth factor I/somatomedin C: a potent inducer of oligodendrocyte development. *Proceedings of the National Academy of Sciences, USA*, **83**, 822–6.

Moses, A.C., Nissley, S.P., Short, P.A., Rechler, M.M., White, R.M., Knight, A.B. & Higa, O.Z. (1980). Increased levels of multiplication-stimulating activity, an insulin-like growth factor, in fetal rat serum. *Proceedings of the National Academy of Sciences, USA*, **77**, 3649–53.

de Pagter-Holthuizen, P., Jansen, M., van Schaik, F.M.A., van der Kammen, R., Oosterwijk, C., Van den Brande, J.L. & Sussenbach, J.S. The human insulin-like growth factor II gene contains two development-specific promoters. *FEBS Letters* (in press).

Rechler, M.M. & Nissley, S.P. (1986). Insulin-like growth factor (IGF)/somatomedin receptor subtypes: structure, function, and relationships to insulin receptors and IGF carrier proteins. *Hormone Research*, **24**, 152–9.

Recio-Pinto, E., Lang, F.F. & Ishii, D.N. (1984). Insulin and insulin-like growth factor II permit nerve growth factor binding and the neurite formation response in cultured human neuroblastoma cells. *Proceedings of the National Academy of Sciences, USA*, **81**, 2562–6.

Rinderknecht, E. & Humbel, R.E. (1978). The amino acid sequence of human insulin-like growth factor I and its structural homology with proinsulin. *The Journal of Biological Chemistry*, **253**, 2769–76.

Roth, R.A., Morgan, D.O., Beaudoin, J. & Sara, V. (1986). Purification and characterization of the human brain insulin receptor. *The Journal of Biological Chemistry*, **261**, 3753–7.

Rotwein, P., Pollock, K.M., Didier, D.K. & Krivi, G.G. (1986). Organization and sequence of the human insulin-like growth factor I gene. *The Journal of Biological Chemistry*, **261**, 4828–32.

Sara, V.R. (1986). Brain Development. In *Mental Retardation and Development Disabilities*, ed. J. Wortis, p.163. Elsevier, New York.

Sara, V.R., Carlsson-Skwirut, C., Andersson, C. Hall, E., Sjögren, B., Holmgren, A. & Jörnvall, H. (1986). Characterization of somatomedins from human fetal brain: identification of a variant form of insulin-like growth factor I. *Proceedings of the National Academy of Sciences, USA*, **83**, 4904–7.

Sara, V.R., Gustavson, K.-H., Annerén, G., Hall, K. & Wetterberg, L. (1983). Somatomedins in Down syndrome. *Biological Psychiatry*, **18**, 803–11.

Sara, V.R. & Hall, K. (1984). The biosynthesis and regulation of fetal somatomedin. In *Fetal Neuroendocrinology*, ed. P. Gluckman & F. Ellendor, p.213, Perinatal Press: New York.

Sara, V.R., Hall, K., Mizaki, M., Fryklund, L., Christensen, N. & Wetterberg, L. (1983). The ontogenesis of somatomedin and insulin receptors on the human fetus. *Journal of Clinical Investigation*, **71**, 1084–94.

Sara, V.R., Hall, K., von Holtz, H., Humbel, R., Sjögren, B. & Wetterberg, L. (1982). Evidence for the presence of specific receptors for insulin-like growth factors I (IGF-I) and II (IGF-II) and insulin throughout the adult human brain. *Neuroscience Letters*, **34**, 39–44.

Sara, V.R., Hall, K. & Wetterberg, L. (1981). Fetal brain growth: proposed model for regulation by embryonic somatomedin. In *The Biology of Normal Human Growth*, ed. M. Ritzén, A. Apéria, K. Hall, A. Larsson & R. Zetterström, p.241. Raven Press, London.

Sara, V.R., King, T.L., Stuart, M.C. & Lazarus, L. (1976). Hormonal regulation of fetal brain cell proliferation: presence in serum of a trophin responsive to growth hormone stimulation. *Endocrinology*, **99**, 90–7.

Sara, V.R. & Lazarus, L. (1974). The prenatal action of growth hormone on brain and behaviour. *Nature*, **250**, 257–8.

Sara, V.R., Lazarus, L., Stuart, M.C. & King, T. (1974). Fetal brain growth; selective action by growth hormone. *Science*, **186**, 446–7.

Sara, V.R., Prisell, P., Enberg, G., and Sjögren, B. (1986). Enhancement of insulin-like growth factor II receptors in glioblastoma. *Cancer Letters*, **32**, 229–234.

Sara, V.R., Uvnäs-Moberg, K., Uvnäs, B., Hall, K., Wetterberg, L., Poslonec, B & Goiny, M. (1982). The distribution of somatomedins in the nervous system of the cat and their release following neural stimulation. *Acta Physiologica Scandinavica*, **115**, 467–70.

Van Schravendijk, C.F.H., Hooghe-Peters, E.L., Van den Brande, J.L. & Pipeleers, D.G. (1986). Receptors for insulin-like growth factors and insulin on murine fetal cortical brain cells. *Biochemical and Biophysical Research Communications*, **135**, 228–38.

Scott, J., Cowell, J., Robertson, M.E., Priestley, L.M., Wadey, R., Hopkins, B., Pritchard, J., Bell, G.I., Rall, L.B., Graham, C.F. & Knott, T.J. (1983). Insulin-like growth factor-II gene expression in Wilms' tumour and embryonic tissues. *Nature*, **317**, 260–2.

Soares, M.B., Ishii, D.N. & Efstratiadis, A. (1985). Developmental and tissue-specific expression of a family of transcripts related to rat insulin-like growth factor II m-RNA. *Nucleic Acids Research*, **13**, 1119–34.

Tham, A., Sara, V.R., Borg, S. & Wetterberg, L. (1986). Circulating levels of insulin-like growth factors 1 and 2 and somatomedin B in alcoholic patients. *Psychiatry Research*, **18**, 301–8.

Ullrich, A., Gray, A., Tam, A.W., Yang-Feng, T., Tsubokawa, M., Collins, C., Henzel, W., Le Bon, T., Kathuria, S., Chen, E., Jacobs, S., Francke, U., Ramachandran, J. & Fujita-Yamaguchi, Y. (1986). Insulin-like growth factor I receptor primary structure: comparison with insulin receptor suggests structural determinants that define functional specificity. *The EMBO Journal*, **5**, 2503–12.

Yang, Y.W.-H., Romanus, J.A., Liu, T.-Y., Nissley, S.P. & Rechler, M.M. (1985). Biosynthesis of rat insulin-like growth factor II. *The Journal of Biological Chemistry*, **260**, 2570–7.

Zumstein, P.P., Lüthi, C. & Humbel, R.E. (1985). Amino acid sequence of a variant pro-form of insulin-like growth factor II. *Proceedings of the National Academy of Sciences, USA*, **82**, 3169–72.

Part IV

Endocrine control of growth and maturation

MICHAEL O. THORNER, MARY LEE VANCE,
ALAN D. ROGOL, ROBERT M.BLIZZARD,
GEORGEANNA KLINGENSMITH,
JENNIFER NAJJAR, CHARLES G.D. BROOK,
PATRICIA SMITH, SEYMOUR REICHLIN,
JEAN RIVIER & WYLIE VALE

Growth hormone and growth hormone releasing hormone

Introduction

Growth hormone (GH) is secreted by the anterior pituitary in a pulsatile fashion. Its secretion is tightly regulated by hypothalamic factors and by feedback from peripheral factors such as serum glucose and fatty acid levels. The hypothalamic input includes the reciprocal secretion of somatostatin and growth hormone releasing hormone (GHRH). Thus, a pulse of GH is mediated by suppression of tonic hypothalamic somatostatin secretion associated with an increase in GHRH secretion. This will not be further discussed but has been reviewed in detail previously (Tannenbaum & Ling, 1984; Thorner *et al* ., 1986). As shown in Figure 1, GH is also regulated by circulating levels of somatomedin C (otherwise known as insulin-like growth factor) which is either produced locally or in the periphery. Thus, somatomedin C inhibits GH secretion at both the pituitary and hypothalamic levels by modulating somatostatin and possibly also GHRH secretion. Another important influence on GH secretion is gonadal steroids. This occurs in both the human and in animals.

Growth hormone secretion during the life cycle in the human

Levels of GH are detectable in the fetus during the mid-trimester and remain high throughout intrauterine life. Detectable GH is found in the serum of human fetuses as early as 70 days of gestation and by mid to late second trimester, values may reach 150 ng/ml. Thereafter, GH levels decline, but at the time of birth and for several weeks thereafter, the levels remain high when compared to adult values. Following delivery, GH levels fall and remain relatively low during childhood and rise again at the time of puberty. Peak secretion occurs during puberty and progressively falls with advancing age; GH secretion is greater in women than in men.

Although the GH radioimmunoassay was one of the first radioimmunoassays to be developed, our knowledge of the pattern of GH secretion during the life cycle in man is rudimentary at the present time. Prior studies were limited by infrequent GH measurements and by lack of objective criteria to assess pulsatile hormone secretion.

Virtually no information exists on spontaneous GH secretion in children less than 6 years of age. In children older than 6 years, GH levels appear to be related to age, sex and pubertal status. There is indirect information, for example, from correlative studies between somatomedin C levels, pubertal stage, and timing of peak growth velocity. Studies are underway in several centres to characterize the pattern of GH secretion in normal children at various stages of development and at various stages of puberty.

Similarly, there are very few comprehensive studies of GH secretion in the adult. In one study we performed, the pattern of GH secretion was characterized in younger

Figure 1. Schematic representation of the control of GH secretion. Growth hormone secretion is regulated by two hypothalamic hormones, somatostatin and GRF; these inhibit and stimulate growth hormone secretion, respectively. The release of GH results in the generation of IGF–1 (somatomedin C) from peripheral issues which produces some of the metabolic effects of growth hormone. IGF–1 exerts a negative feedback directly at the pituitary and also hypothalamus by stimulating somatostatin secretion and possibly by inhibiting GHRH secretion. Note that IGF–1 cannot be generated in the absence of adequate nutrition. (Reproduced with kind permission from Thorner *et al.*, 1986).

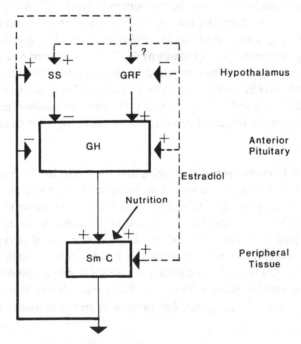

man (18–30 years), younger women (18–33 years in the early follicular phase of the menstrual cycle), older men (56–71 years), and older women (57–76 years; postmenopausal) (Ho *et al.*, 1987). All individuals were healthy and none were taking any medication. Growth hormone concentrations were measured in blood samples obtained at 20 minute intervals during the 24 hour study. These studies demonstrated that age and sex have independent and interrelated effects on both total GH secretion and on certain aspects of its pulsatile secretion. Younger, but not older, women secrete significantly more GH than do their age matched male counterparts as was predicted by earlier studies. Thus, while there is no obvious age related decline in GH secretion in men, such diminished release clearly occurs in women. Consistent with these observations, estradiol but not testosterone, strongly correlated with both total GH secretion and the fraction of GH secreted in pulses.

Nutritional factors in growth hormone secretion

It has been recognized for many years that children with systemic illness or who have a malabsorption syndrome or are malnourished do not grow. Additionally, children with protein calorie malnutrition have elevated levels of GH but do not grow (Pimstone, Wittman & Hansen, 1966; Soliman *et al.*, 1986). Several years ago Clemmons and colleagues demonstrated that normal subjects who were fasted for 8 days had a fall in circulating levels of somatomedin C(Clemmons, Klibanski & Underwood, 1981). Since somatomedin C exerts a negative feedback on GH secretion at the level of the pituitary and hypothalamus, the very low levels of GH (often undetectable) in the circulation of normal adults may be accounted for by the "high" somatomedin C levels which occur as a result of excessive nutrition. In addition, a number of unpredictable factors including food intake, stress, and sleep all may affect GH secretion. Thus, the 24 hour pulsatile pattern of GH secretion obtained from repetitive venous sampling in normal men during a control fed day and during a 5 day fast may uncover an ultradian periodicity of GH secretion. We have performed this study and demonstrated that there was approximately a four-fold increase in integrated GH concentration (2.8 ± 0.5 vs. 8.8 ± 0.8 µg/min/ml) which was predominantly caused by an increase in pulse frequency from 5.8 ± 0.7 to 9.9 ± 0.7 pulses per 24 hours with only a minor change in pulse amplitude (Ho *et al.*, 1986). Additionally, there was an increase in the interpulse GH concentration during the fast. Time series analysis revealed the emergence of several GH rhythms and one with a periodicity of 91 minutes was discernable only during the first and fifth fast days and not in the fed state. We believe that the nutritional state is an important determinant of pulsatile GH secretion. In addition, GH is unlikely to stimulate linear growth unless there is sufficient nutrition. Growth hormone is only able to generate somatomedin C in the presence of adequate caloric intake with the appropriate proportion of protein, carbohydrate, and fat. Nutrition plays a vital role in the

Figure 2. GH release in response to hpGRF–40 (human pancreatic growth hormone releasing hormone–40) in children with short stature. The horizontal axis is time (minutes) before and after the iv injection of hGRH–40, 3.3µg/kg as a bolus dose. The vertical axis is GH concentration (nanograms per ml). Each set of symbols represents an individual patient. A, IGHD; B, organic hypopituitarism; C, IUGR; D, CD and/or familial short stature. Bars at the lower right of each panel are the mean ± SE of the peak responses to the arginine/L-dopa and hpGRH–40 stimulation tests. (Reproduced with kind permission from Rogol *et al.*, 1984).

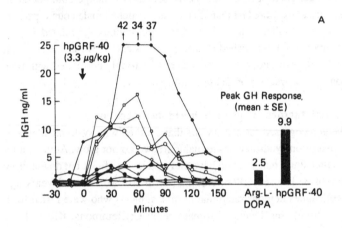

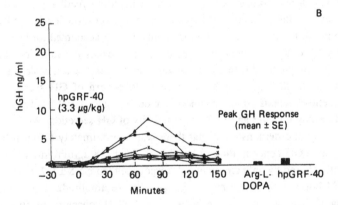

regulation of growth and development. Hormonal manipulation alone, without attention to nutritional aspects, is unlikely to promote optimal growth.

Growth hormone deficiency

A small proportion of children with short stature suffer from GH deficiency. The etiology of short stature is often unknown but may result from either an organic lesion, usually a hypothalamic tumor, or more commonly, is of undetermined etiology (idiopathic). Growth hormone deficiency rarely results from a defect in the GH gene such that normal GH is not produced. With the exception of this syndrome, the vast majority of growth hormone deficient children suffer from

Figure 2 (*contd.*)

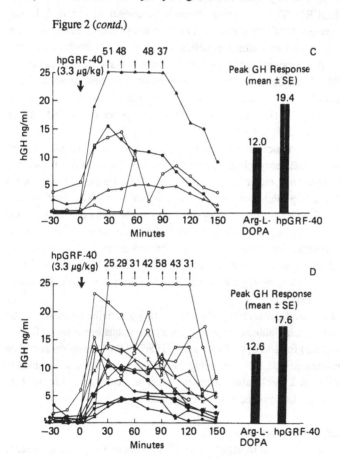

Table 1. *Results of GHRH treatment in 24 GH-deficient children*

Therapy	GHRH dose µg/kg/day	Growth velocity (mean ± S.D.; cm/year)	
		pretreatment	at 6 months
Every 3h	19	3.5 ± 1.4	10.0 ± 2.2
Every 3h overnight	6–8	3.6 ± 1.0	6.2 ± 2.1
b.i.d.	8	3.2 ± 1.8	7.9 ± 2.4

dysregulation of the pituitary since these children have an increase in GH after administration of GHRH. Thus, the somatotrope is functional. In Figure 2 the GH responses to exogenous GRF are shown in children with various causes of short stature including those with idiopathic GH deficiency. Note that in this group there is great variability in the response, but almost all the children have an increase in GH (Rogol et al., 1984). These observations have been reported by several other groups and it is now generally accepted that it is possible to stimulate GH secretion in GH deficient children with GHRH. It, therefore, seemed appropriate that GHRH should be investigated as a therapeutic agent.

We have performed a multicentre study using GHRH [GHRH(1–40)–OH] to treat 24 growth hormone deficienty children for at least 6 months. GHRH–40 was administered subcutaneously, either by twice daily injections, 4 µg/kg per dose, or via peristaltic portable pump (Pulsamat, by courtesy of Dr D. Linkie, Ferring Inc., Ridgewood, NJ), 1–3 µg/kg per dose, every 3 hours over 24 hours or every 3 hours via pump at night only.Twenty-two of 24 children had an increase in growth velocity during GHRH treatment. The mean (± SD) growth velocities before and during treatment by the 3 regimens are summarized in Table 1. Six children received both pump and twice daily treatment regimens sequentially. The growth velocities were similar for both treatments.

To determine the potential feasibility of administering a sustained release GHRH preparation, 5 normal male volunteers were given a continuous intravenous GHRH infusion (10 ng/kg/min) for 14 days (Vance et al., 1986). All of the subjects had an increase in serum somatomedin C levels during the 14 day infusion; these declined to pretreatment levels over 2 weeks after discontinuation of the infusion. On the 14th day of infusion, growth hormone secretion was pulsatile.

Summary and conclusion

These preliminary data indicate that GHRH produces a significant increase in growth rate in GH deficient children when administered by different treatment schedules. The results of 14 days of continuous somatotrope stimulation by GHRH–40 suggest that a sustained release GHRH (or analog) preparation may be a feasible method of administering this hormone. It appears that this form of treatment will result in restoration of pulsatile growth hormone secretion and growth in GH deficient children.

References

Clemmons, D.R., Klibanski, A. & Underwood, L. (1981). Reduction of plasma immunoreactive somatomedin–C during fasting in humans. *Journal of Clinical Endocrinology and Metabolism*, **53**, 1247–50.

Ho, K.Y., Evans, W.S., Blizzard, R.M., Veldhuis, J.D., Merriam, G.R., Samojilik, E., Furlanetto, R., Rogol, A.D., Kaiser, D.L. & Thorner, M.O. (1987). Effects of sex and age on the 24 hour secretory profile of GH secretion in man: Importance of endogenous estradiol concentrations. *Journal of Clinical Endocrinology and Metabolism*, **64**, 51–8.

Ho, K.Y., Furlanetto, R., Alberti, K.G.M.M. & Thorner, M.O. (1986). Fasting unmasks an intrinsic pulsatile patern of GH secretion in man. *68th Annual Meeting of the Endocrine Society*, Anaheim, California.

Pimstone, B.L., Wittman, W. & Hansen, J.D.L. (1966). Growth hormone and kwashiorkor. *Lancet*, **2**, 779–80.

Rogol, A.D., Blizzard, R.M., Johanson, A.J., Furlanetto, R.W., Evans, W.S., Rivier, J., Vale, W.W. & Thorner, M.O. (1984). Growth hormone release in response to human pancreatic tumor growth hormone releasing factor–40 in children with short stature. *Journal of Clinical Endocrinology and Metabolism*, **59**, 580–6.

Soliman, A.T., Hassan, A.E.H.I., Aref, M.K., Hintz, R.L., Rosenfeld, R.G. & Rogol, A.D. (1986). Serum insulin-like growth factors I and II concentrations and growth hormone and insulin responses to arginine infusion in children with protein-energy malnutrition before and after nutritional rehabailitation. *Pediatric Research*, **20**, 1122–30.

Tannenbaum, G.S. & Ling, N. (1984). The interrelationship of growth hormone (GH)-releasing factor and somatostatin in generation of the ultradian rhythm of GH secretion. *Endocrinology*, **115**, 1952–7.

Thorner, M.O., Vance, M.L.,Evans, W.S., Blizzard, R.M., Rogol, A.D., Ho, K., Leong, D.A., Borges, J.L.C., Cronin, M.J., MacLeod, R.M., Kovacs, K., Asa, S., Horvath, E., Frohman, L., Furlanetto, R., Klingensmith, G., Brook, C., Smith, P., Reichlin, S., Rivier, J. & Vale, W. (1986). Physiological and clinical studies of GRF and GH. *Recent Progress in Hormone Research*, **42**, 589–640.

Vance, M.L., Evans, W.S., & Thorner, M.O. (1986). Growth hormone secretion is augmented during 14 days of continuous growth hormone releasing infusion in normal man. *Clinical Research*, **34**, 435A.

R. STANHOPE

The endocrine control of puberty

Introduction

Schally isolated and sequenced the decapeptide gonadotrophin releasing hormone (GnRH) in 1971 and demonstrated that it was responsible for the control of the release of two hormones from the pituitary gland: luteinising hormone (LH) and follicle-stimulating hormone (FSH). The mechanism whereby GnRH controls the release of these two gonadotrophins will be the subject of this article.

It has long been realised that anterior pituitary hormones are secreted in diurnal rhythms, but only during the last 15 years has the pulsatile nature of the secretion of these hormones been appreciated. It is the pulsatile release and the modulation of pulse frequency and amplitude, that is fundamental to their biological action. However, the continuous basal secretion of these hormones is probably also significant.

The GnRH gene

Recent experiments in the hypogonadal mouse have revealed that this animal has a 33 kilobase deletional mutation which involves the GnRH gene (Masin et al., 1986). This GnRH gene has been synthesised and replaced into hypogonadotrophic mouse embryos (Mason et al., 1986). Such transgenic mice have full reproductive capability and GnRH expression restored in the appropriate hypothalamic neurones. The unusual feature of this model is that it does not appear to be important where the GnRH gene is replaced in the mouse genome. The clinical parallel of Kallman syndrome to the hypogonadal mouse model is interesting, and may provide a new potential treatment modality.

The human GnRH gene has now been sequenced (Nikolics et al., 1985). As well as a sequence coding for the decapeptide GnRH, there is an area coding for a 56 amino acid peptide, Gonadotrophin Associated Peptide (GAP). This peptide is synthesised in an equimolar proportion to GnRH and stimulates the release of gonadotrophins in rat pituitary cell cultures. GAP has also been described as a prolactin inhibitor which would explain the differential responses of prolactin secretion to endogenous GnRH secretion and to exogenously administered GnRH (administered in the absence of GAP). However, much further work will be required to substantiate this hypothesis.

The pulsatile nature of GnRH secretion

We owe much to Knobil's experiments (Knobil, 1980) on the rhesus monkey for our understanding of the significance of pulsatile GnRH secretion. This provided the fundamental model on which many of the modern treatment regimens in reproductive endocrinology are based. Rhesus monkeys, following an artificial lesion induced in mediobasal hypothalamus, have no endogenous GnRH secretion. The effect of GnRH on gonadotrophin secretion and gonadal maturation was observed by the administration of native GnRH via a 'hypothalamic prosthesis'. In a series of now classical experiments, Knobil found that pulsatile, and not continuous, administration of GnRH was essential for gonadal maturation. Indeed, pulse frequency was critical. An elegant experiment (Caraty, Martin & Montgomery, 1984) in ewes has shown that the administration of GnRH antibodies could abolish the pulsatile nature of gonadotrophin secretion but the subsequent pulsatile administration of GnRH analogue (not recognised by antibodies to native GnRH) could restore pulsatile gonadotrophin secretion.

A further important experiment about the nature and timing of puberty was performed by Knobil (1980) in prepubertal rhesus monkeys. The administration of pulsatile GnRH at an appropriate frequency permitted gonadal maturation and sexual maturity. However, on withdrawal of pulsatile GnRH treatment, the monkeys returned to their prepubertal state. This did not affect the pattern and timing of the monkeys' own endogenous puberty which occurred at the expected time. Thus, puberty in the rhesus monkey is GnRH dependent and exogenously administered GnRH did not affect the monkeys' own hypothalamic 'time clock'.

Gonadal maturation during childhood

We know more of gonadal maturation in girls than of boys because of the use of pelvic ultrasound (Stanhope et al., 1985). This is a non-invasive procedure and can be performed in normal girls as well as girls with abnormalities of sexual maturation. The findings of pelvic ultrasound in normal girls have been confirmed by post-mortem studies (Peters, Byskov & Grinster, 1981). Ovarian maturation continues throughout childhood and is not confined to the peripubertal period. There is a gradual increase in both the size of the ovaries and the number and size of follicular cysts throughout childhood. We believe these appearances to be due to pulsatile gonadotrophin secretion and certainly anencephalic fetuses have ovaries without cysts. The ovarian morphological appearance which is characteristic of pulsatile gonadotrophic secretion, the 'multicystic' ovary, is seen using ultrasound in normal girls from approximately eight years of age. Using our knowledge of ovarian maturation and gonadotrophin pulsatility in studies of delayed puberty we have hypothesised that there must be a progressive increase in GnRH secretion during childhood.

The endocrinology of human puberty

For ethical considerations we have relied on experiments in normal children performed in the early 1970s. The first endocrine event of puberty is an increase in the amplitude of pulsatile LH secretion at night (Boyar *et al.*, 1972). During early puberty the amplitude of these nocturnal LH pulses increases and towards the latter half of puberty pulsatile LH secretion occurs in high amplitude during the day as well as at night. There is a different pattern of sex steroid secretion between the sexes. This has important implications for the timing of blood samples for sex steroid assays. In boys, testosterone starts to rise about 60 to 90 minutes after the first high amplitude LH pulse of the night (Boyar *et al.*, 1974) whereas in girls the rise in serum oestradiol occurs much later in the night and rises to a peak at mid-morning (Boyar *et al.*, 1976). Such studies emphasise the importance both of studying pulsatile secretion of pituitary hormones at night as well as the significance of the night in sexual maturation. The importance of the night in relation to the endocrinology of reproduction is not confined to puberty; the onset of the pre-ovulatory LH surge in women usually occurs during the early hours of the morning.

Studies in central precocious puberty

The pattern of spontaneous gonadotrophin secretion in girls with central precocious puberty is similar to that of normal puberty. If girls with central precocious puberty are treated with subcutaneous (D–Trp[6]) GnRH analogue, suppression of gonadotrophin secretion occurs with subsequent regression of the clinical signs of puberty (Crowley *et al.*, 1981). Administration of the less potent (D–Ser[6]) GnRH analogue intranasally does not suppress total gonadotrophin secretion but abolishes gonadotrophin pulsatility (Stanhope, Adams & Brook, 1985a). This results in suppression of the ovarian multicystic morphology and consequently either arrest or regression of the signs of precocious puberty. Persistence of GnRH analogue therapy leads to suppression of ovarian size into the normal range for chronological age, despite an increase in basal gonadotrophin concentrations.

Isolated premature thelarche

A variant of premature sexual maturation exists in which only isolated breast development occurs, usually in a cyclical pattern, without other signs of puberty. It has been observed (Mills, 1985) that in this condition isolated ovarian cysts are commonly seen and that serum FSH (Follicle Stimulating Hormone) levels are greater than LH. Certainly pelvic ultrasound and studies of gonadotrophin pulsatility distinguish between isolated premature thelarche and central precocious puberty (Stanhope *et al.*, 1986). It is probable that high levels of secretion of FSH cause growth of isolated or small numbers of large ovarian cysts which secrete oestrogen; in the absence of high concentrations of LH, these cysts do not luteinise. The clinical result is breast development which often waxes and wanes, in parallel to the size of ovarian cysts, without associated androgen secretion which results in absence of pubic hair and normal skeletal and linear growth.

Recent understanding of the inhibin group of molecules (Ling *et al.*, 1986; Vale *et al.*, 1986) have revealed a dual method for the control of gonadotrophin secretion. Certainly these hormones do not act via GnRH receptors on the pituitary gonadotrope. It may be that isolated premature thelarche is caused by an abnormality of inhibin/activin biosynthesis or secretion. Contrary to the paediatric literature, it is probable that some girls with isolated premature thelarche retain a bias for isolated ovarian cyst formation in adult life (Brook *et al.*, 1987) which may result in infertility.

Pulsatile GnRH treatment of delayed puberty

The administration of pulsatile GnRH has been shown to restore fertility in both men (Hoffman & Crowley, 1982; Morris *et al.*, 1984) and women (Crowley & McArthur, 1980; Mason *et al.*, 1984) with hypogonadotrophic hypogonadism. However, if doses used for the treatment of adults with infertility are used in children with hypogonadotrophic hypogonadism then either inappropriately accelerated pubertal maturation or pituitary desensitisation results (Stanhope *et al.*, 1985). However, pulsatile GnRH therapy can be given in a way that mimics normal physiology; during the first half of puberty pulsatile GnRH therapy is given only at night with a gradual increase in pulse amplitude in order to induce progression in sexual maturation. During the second half of puberty, pulsatile GnRH therapy is administered during the day as well as at night. Using such a regimen all the clinical features, growth acceleration, ovarian maturation and endocrinology of normal puberty has been mimicked (Stanhope *et al.*, 1987). Certainly the pattern of pulsatile gonadotrophin secretion at night with its associated characteristics of sex steroid secretion is almost identical to that of normal puberty (Stanhope, Adams & Brook, 1985b). The only difference between puberty induced by pulsatile GnRH therapy and normal puberty is that menarche is more frequently ovulatory than would be expected. This is probably an artefact of the treatment regimen of administering pulsatile GnRH throughout the 24 hours.

Mechanism of the growth acceleration of puberty

Although growth hormone (GH) and sex steroids are synergistic during the growth acceleration of puberty, there is a dissociation of sex steroids secretion and the onset of the growth spurt. In pubertal boys serum testosterone concentration rises progressively and yet the onset of the growth acceleration only occurs between genitalia stages 3 and 4. Clearly the onset of the growth acceleration is not entirely sex steroid dependent.

In girls treatment with pulsatile GnRH, there is a dramatic increase in GH pulse amplitude within a week of commencing treatment (Stanhope, Pringle & Brook, 1985); the growth acceleration of normal girls commences at the onset of breast development. In contrast, boys treated with pulsatile GnRH commence their growth spurt at the expected time, between genitalia stages 3 and 4. Initially, during early puberty, GH secretion decreases as growth decelerates. At the attainment of genitalia stage 3 to 4, there is an increase in GH pulse amplitude coincident with the onset of the growth spurt (Brook *et al.*, 1987). When peak height velocity is achieved at genitalia stage 4 to 5, GH pulse amplitude attains its peak. In both girls and boys there was no variation in the frequency of GH secretion; alteration of GH secretion was entirely by amplitude modulation. As the growth acceleration induced by pulsatile GnRH therapy is identical to that of normal puberty, it is probable that these findings of altered GH secretory dynamics are relevant to the mechanism of the growth acceleration of normal girls and boys.

Pulse frequency of GnRH secretion

The pulse frequency of gonadotrophin secretion during puberty is approximately two hourly (Jakacki *et al.*, 1982). However, we know much more of the significance of pulse frequency modulation in adults than we do in adolescents. Certainly the use of a 90 minute pulse frequency of GnRH administration can mimic the endocrinology of puberty (Stanhope *et al.*, 1985b) and the normal menstrual cycle (Masin *et al.*, 1984). There are characteristic pulse frequency changes of gonadotrophin secretion during various phases of the normal menstrual cycle (Backstrom *et al.*, 1982) although these may not be as pronounced as was previously believed (Reame *et al.*, 1984). Certainly Knobil (1980) in the rhesus monkey model had demonstrated that the pulse frequency of GnRH secretion could determine the ratio of LH to FSH secretion. High frequency GnRH secretion, such as that of the follicular phase of the menstrual cycle, is associated with high FHS:LH ratios; the reverse situation occurs in the luteal phase. The importance of such control mechanisms during puberty will require further study.

The control of GnRH secretion

GnRH secretion can potentially be modified by gonadotrophins (short-loop feedback) or by sex steroids (long-loop feedback). The short-loop feedback has little effect on GnRH secretion in the rhesus monkey (Kesner *et al.*, 1986). However, the contrary is true of modification of the GnRH pulse generator by sex steroid secretion. If male rhesus monkeys are castrated, there is a rise of gonadotrophin baseline secretion as well as a marked increase in gonadotrophin pulse frequency as well as pulsatility. An analogous phenomenon is seen in girls with the Turner syndrome after the administration of oestrogen.

Hypothalamo–pituitary–gonadal axis

A schematic diagram of the hypothalamo–pituitary–gonadal axis is shown in Figure 1. Like many biological control systems this is a cascade with the hypothalamus being the interface between the neurological and endocrine systems. The predominant amplification of pulses occurs at the pituitary whereas frequency modulation occurs at the level of the hypothalamus. The prolongation of the pulse signal, from milliseconds, to seconds, to minutes occurs during the passage of the cascade. Although Knobil and colleagues (Wilson *et al.*, 1984) have demonstrated that electrical activity in the region of the median eminence correlates with the release of GnRH pulses, this is only multi-unit activity (MUA) and does not represent the site of the pulse generator. GnRH synthesising neurons are found in several parts of the hypothalamus but they have many interconnections throughout the central nervous system. The exact site of the GnRH 'pulse generator' is, as yet, unknown.

Conclusion

We are now able to switch pubertal maturation on and off in a physiological pattern and this has enabled us to learn much of the physiology of normal puberty. Our knowledge of gonadal maturation during childhood has led to the belief that there is no initiator of puberty. There is a gradual maturation of the hypothalamo-pituitary-gonadal axis during childhood. The progressive increase in periodicity and

Figure 1. The role of the hypothalamus, pituitary gland and gonads in modulating the biological signal of the GnRH 'pulse generator'.

Hypothalamic - Pituitary - Gonadal Axis

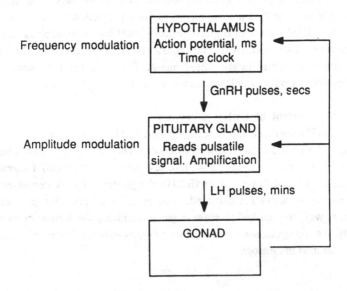

amplitude of GnRH secretion results in gonadal maturation and at a time when sufficient sex steroid secretion is produced to initiate secondary sexual characteristics, then puberty begins.

References

Backstrom, C.T., McNeilly, A.S., Leask, R.M. & Baird, D.T. (1982). Pulsatile secretion of LH, FSH, Prolactin, Oestradiol and Progesterone during the human menstrual cycle. *Clinical Endocrinology*, **17**, 29–42.

Boyar, R.M., Finkelstein, R., Roffwarg, H., Kapen, S., Weitzman, E.D. & Hellman, L. (1972). Synchronization of augmented luteinising hormone secretion with sleep during puberty. *New England Journal of Medicine*, **287**, 582–6.

Boyar, R.M., Rosenfield, R.S., Kapen, S., Finkelstein, J.W., Forrwarg, H.P., Weitzman, E.D. & Hellman, L. (1974). Human puberty – simultaneous augmented secretion of luteinising hormone and testosterone during sleep. *Journal of Clinical Investigation*, **54**, 609–14.

Boyar, R.M., Wu, R.H.F., Roffwarg, H., Kapen, S., Weitzman, E.D., Hellman, L. & Finkelstein, J.W. (1976). Human puberty: 24 hr estradiol patterns in pubertal girls. *Journal of Clinical Endocrinology and Metabolism*, **43**, 1418–21.

Brook, C.G.D., Jacobs, H.S., Stanhope, R., Adams, J. & Hindmarsh, P. (1987). Pulsatility of reproductive hormones: applications to the understanding of puberty and to the treatment of infertility. *Baillières Clinical Endocrinology and Metabolism*, **1**, 23–41.

Caraty, A., Martin, G.B. & Montgomery, G. (1984). A new method for studying pituitary responsiveness in vivo using pulses of LH–RH analogue in ewes passively immunised against native LH–RH. *Reproduction, Nutrition, Development*, **24**, 439–48.

Crowley, W.F., Comite, F., Vale, W.W., Rivier, J., Loriaux, D.L. & Cutler, G.B. (1981). Therapeutic use of pituitary desensitization with a long acting LHRH agonist: A potential new treatment for idiopathic precocious puberty. *Journal of Clinical Endocrinology and Metabolism*, **52**, 370–2.

Crowley, W.F. & McArthur, J.W. (1980). Simulation of the normal menstrual cycle in Kallman's syndrome by pulsatile administration of luteinising hormone releasing hormone (LHRH). *Journal of Clinical Endocrinology and Metabolism*, **51**, 173–5.

Hoffman, A.D.R. & Crowley, W.F. (1982). Induction of puberty in men by long term pulsatile administration of low dose gonadotrophin releasing hormone. *New England Journal of Medicine*, **307**, 1237–41.

Jakacki, R.J., Kelch, R.P., Sauder, S.E., Lloyd, J.S., Hopwood, N.J. & Marshall, J.C. (1982). Pulsatile secretion of luteinising hormone in children. *Journal of Clinical Endocrinology and Metabolism*, **55**, 453–8.

Kesner, J.S., Kaufman, J.M., Wilson, R.C., Kuroda, G. & Knobil, E. (1986). On the short-loop feedback regulation of the hypothalamic luteinising hormone releasing hormone "pulse generator" in the rhesus monkey. *Neuroendocrinology*, **42**, 109–11.

Knobil, E. (1980). The neuro-endocrine control of the menstrual cycle. *Recent Progress in Hormone Research*, **36**, 53–88.

Ling, N., Ying, S–H., Ueno, N., Shimasaki, S., Esch, F., Hotta, M. & Guillemin, R. (1986). Pituitary FSH is released by a heterodimer of the B-subunits from the two forms of inhibin. *Nature*, **321**, 779–82.

Mason, P., Adams, J., Morris, D.V., Tucker, M., Price, J., Voulgaris, Z., Van Der Spuy, Z.M., Sutherland, I., Chambers, G.R., White, S., Wheeler, M.J. &

Jacobs, H.S. (1984). The induction of ovulation with pulsatile luteinising hormone releasing hormone. *British Medical Journal*, **288**, 181–5.

Mason, A.J., Hayflick, J.S., Zoeller, R.T., Young, W.S., Phillips, H.S., Nikolics, K. & Seeburg, P.H. (1986). A deletion truncating the gonadotrophin-releasing hormone gene is responsible for hypogonadism in the hpg mouse. *Science*, **234**, 1366–71.

Masin, A.J., Pitts, F.L., Nikolics, K., Szonyi, E., Wilcox, J.N., Seeburg, P.H. & Stewart, T.A. (1986). The hypogonadal mouse: reproductive functions restored by gene therapy. *Science*, **234**, 1372–8.

Mills, J.L. (1985). Endocrinology of premature thelarche. In *Estrogens in the Environment (ii): Influences on Development*, ed. J.A. McLachlan, pp. 412–27. Elsevier, New York.

Morris, D.V., Adeniyi–Jones, M., Wheeler, M., Sonksen, P. & Jacobs, H.S. (1984). The treatment of hypogonadotrophic hypogonadism in men by the pulsatile infusion of luteinising hormone-releasing hormone. *Clinical Endocrinology*, **21**, 189–200.

Nikolics, K., Masin, A.J., Szonyi, E., Ramachadran, J. & Seeburg, P.H. (1985). A prolactin-inhibiting factor within the precursor for human gonadotrophin-releasing hormone. *Nature*, **316**, 511–17.

Peters, H., Byskov, A.G. & Grinster, J. (1981). The development of the ovary during childhood in health and disease. In *Functional Morphology of the Human Ovary*, ed. J.R.T. Coutts, pp. 26–34. MPT Press, Lancaster.

Plant, T.M. (1983). Pulsatile LH secretion in the male rhesus monkey (Macaca Mulatta): An index of the activity of the hypothalamic GnRH pulse generator. In *Brain and Pituitary Peptides, Ferring Symposium*, pp. 125–39. Karger, Basel.

Reame, M., Sauter, S.E., Kelch, P. & Marshall, J.C. (1984). Pulsatile gonadotrophin secretion during the human menstrual cycle: evidence for altered frequency of gonadotrophin releasing hormone secretion. *Journal of Clinical Endocrinology and Metabolism*, **59**, 328–37.

Schally, A.V., Arimuria, A., Kastin, A.J., Matsuo, H., Baba, Y., Redding, T.W., Nair, R.M.G., Debeljuk, L. & White, W.F. (1971). Gonadotrophin-releasing hormone: One polypeptide regulates secretion of luteinising and follicle-stimulating hormones. *Science*, **173**, 1036–8.

Stanhope, R., Adams, J., Jacobs, H.S. & Brook, C.G.D. (1985). Ovarian ultrasound assessment in normal children, idiopathic precocious puberty and during low dose pulsatile GnRH therapy of hypogonadotrophic hypogonadism. *Archives of Disease in Childhood*, **60**, 116–9.

Stanhope, R., Pringle, P.J. & Brook, C.G.D. (1985). Alteration in the nocturnal pulsatile relase of GH during the induction of puberty using low dose pulsatile LHRH. *Clinical Endocrinology*, **22**, 117–20.

Stanhope, R., Adams, J. & Brook, C.G.D. (1985a). The treatment of central precocious puberty using an intranasal LHRH analogue (Buserelin). *Clinical Endocrinology*, **22**, 795–806.

Stanhope, R., Adams, J. & Brook, C.G.D. (1985b). Disturbances of puberty. *Clinics in Obstetrics and Gynaecology*, **12**, 557–77.

Stanhope, R., Abdulwahid, N.A., Adams, J., Jacobs, H.S. & Brook, C.G.D. (1985). Problems in the use of pulsatile gonadotrophin-releasing hormone for the induction of puberty. *Hormone Research*, **22**, 74–7.

Stanhope, R., Abdulwahaid, N.A., Adams, J. & Brook, C.G.D. (1986). Studies of gonadotrophin pulsatility and pelvic ultrasound examinations distinguish between isolated premature thelarche and central precocious puberty. *European Journal of Pediatrics*, **145**, 190–4.

Stanhope, R., Pringle, P.J., Adams, J., Jacobs, H.S. & Brook, C.G.D. (1987). The endocrine events of puberty elucidated by low dose pulsatile GnRH therapy. *Journal of Endocrinology*, **112** suppl. 34.

Vale, W., Rivier, J., Vaughan, J., McClintock, R., Corrigan, A., Woo, W., Karr, D. & Spiess, J. (1986). Purification and Characterization of an FSH releasing protein from porcine ovarian follicular fluid. *Nature*, **321**, 776–9.

Wilson, R.C., Kesner, J.S., Kaufman, J.M., Uemura, T., Akema, T. & Knobil, E. (1984). Central electrophysiologic correlates of pulsatile luteinising hormone secretion in the rhesus monkey. *Neuroendocrinology* , *39*, 256–60.

J. MÜLLER, C. THØGER NIELSEN &
N.E. SKAKKEBÆK

Testicular maturation, and pubertal growth and development in normal boys

Introduction

The testis is an active organ as early as the 7th week of gestation (for review cf. Faiman, Winter & Reyes, 1981). After a period of low activity during the second part of intrauterine life, the hypothalamic-hypophyseal-gonadal axis is reactivated shortly after birth (Winter & Faiman, 1972; Forest, Cathiard & Bertrand, 1973). After a short, postnatal period with high serum levels of gonadotropins and testosterone, hormone production in the testis decreases and remains low until puberty. Thus, the testis plays an important endocrine role many years before puberty. It is the scope of the present paper to review aspects of early testicular maturation as well as the relationship between the endocrine activity of the testis, initiation of sperm production, and growth and development of the body.

The neonatal period

During the first 3 months of life an increase in serum levels of luteinizing hormone (LH) and follicle stimulating hormone (FSH) occurs (Winter & Faiman, 1972). This is accompanied by a rise in serum testosterone to a mean of approximately 200 ng/100 ml (Forest et al., 1973) a level which is in the range of the adult male. After the age of 3 months serum levels of LH, FSH, and testosterone fall to prepubertal values (serum testosterone less than 6 ng/100 ml). Recently, it has been demonstrated that this reactivation of the hypothalamic-hypophyseal-gonadal axis is associated with an increase and subsequent decrease of testicular size (Cassorla et al., 1981; Siebert, 1982). The increased endocrine activity also seems to affect the development of the germ cells and Sertoli cells during the neonatal period. In a study of 46 pairs of testes from normal boys who died suddenly and unexpectedly, we found a peak in the number of germ cells between day 50 and day 150 after birth (Müller & Skakkebæk, 1984) (Figure 1). In addition, it has been shown that the number of Sertoli cells increases from 130×10^6 per testis to 750×10^6 during infancy (Cortes, Müller & Skakkebæk, 1987). The significance of these observations has not yet been elucidated. It is interesting, however, that although adult levels of gonadotropins and testosterone are reached during the first 3 months of life, this is not accompanied by maturation of the germ cells and Sertoli cells nor development of secondary sex characteristics of the infant. Some boys with undescended testes have been shown to have a diminished neonatal peak of LH, FHS and testosterone (Gendrel, Roger & Job, 1980). Furthermore, cryptorchidism is associated with decreased fertility in adulthood. These findings are in line with the hypothesis that the early endocrine activity is of importance for the subsequent maturation and development of the testis.

Childhood

Despite the fact that Leydig cells are not apparent on light microscopy during childhood, a significant production of testosterone from the testis has been demonstrated by means of testicular vein catheterization (Forti *et al.*, 1981) and analysis of peritesticular fluid (Karpe *et al.*, 1982). Furthermore, we have demonstrated that the size of the testis increases from a median of 0.6 ml to a median of 1.6 ml during the 0–10 year period (Müller & Skakkebæk, 1983). These latter values were obtained by weighing gonads from boys who died suddenly. The total number of germ cells per testis determined by stereological analysis of the gonads increased from 6×10^6 to 42×10^6 during childhood. In addition, the germ cells underwent some maturation as the size of the nuclei gradually decreased (Müller & Skakkebæk, 1983), and the location of the cells within the seminiferous tubules

Figure 1. Relationship between changes in the number of germ cells in the testis (●), serum levels of FSH (O); (adapted from Winter & Faiman, 1972), and serum levels of testosterone (▲); (adapted from Forest *et al.*, 1973) during the neonatal period. Reproduced with permission from Müller & Skakkebæk, (1984).

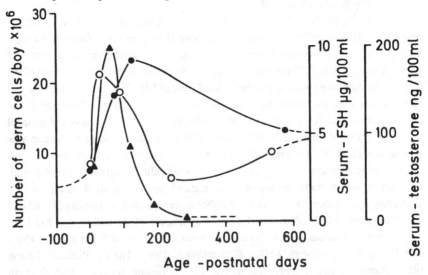

changed from preferentially central to a more peripheral one (Müller & Gundersen, unpublished data).

These observations support the concept that the testis is an active organ before puberty. However, the significance of this activity in relation to growth and development still remains to be investigated.

Puberty

It is well established that androgens play an important role in pubertal growth and development (Aynsley-Green, Zachmann & Prader, 1976; Tanner *et al.*, 1976). Until recently male puberty has been described in relation to changes in height, secondary sex characteristics, and serum levels of different hormones. However, the very important question of maturation of the seminiferous epithelium in relation to other pubertal events has been difficult to address. It has been known for a long time that postpubertal boys excrete spermatozoa in the urine (Baldwin, 1928; Richardson & Short, 1978; Hirsch *et al.*, 1985). By means of analysis of urine samples from pubertal boys, we have studied the relationship between the onset of release of spermatozoa (spermarche), and the development of pubic hair, growth of the testes, peak height velocity, sitting height and testosterone excretion in the urine (Nielsen *et al.*, 1986b; Nielsen *et al.*, 1986a).

Forty Scottish boys, 8.6–11.7 years old, were followed with 24-hour urine samples every 3 months and physical examinations at 6-monthly intervals until the age of 12.0–18.3 years. After centrifugation, the urine samples were analysed for the presence of spermatozoa, and spermarche was estimated on the basis of age at first observed spermaturia (Nielsen *et al.*, 1986a). Peak height velocity was calculated as described by Preece and Baines (1978), and urinary testosterone was determined as previously described (Neilsen *et al.*, 1986b).

Figure 2. Age at spermarche, median 13.4 years. Reproduced with permission from Nielsen *et al.*, (1986a).

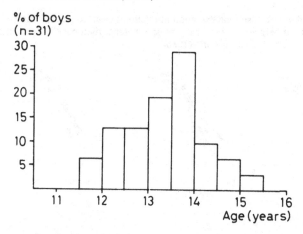

The median age at spermarche was 13.4 years (Figure 2), and the median pubic hair stage (Tanner) at spermarche was 2.5 (Figure 3). More than 50% of the boys had spermarche when only in stage 2 of puberty, and a single individual was entirely prepubertal as judged from the development of secondary sex characteristics. Thus, spermarche was found to be an early event of male puberty.

Spermarche seems to be closely related to peak height velocity and preceeded this phenomenon in most boys (Nielsen *et al.*, 1986a). Furthermore, the peak velocity of

Figure 3. Estimated pubic hair stage at spermarche. Median 2.5. Reproduced with permission from Nielsen *et al.*, (1986a).

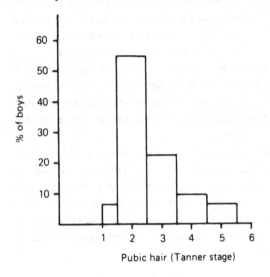

Figure 4. Urinary testosterone excretion in relation to age (left panel), time from peak velocity of height (centre panel), and spermarche (right panel). Reproduced with permission from Nielsen *et al.*, (1986b).

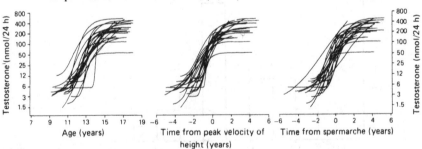

leg length (median 13.7 years) preceeded peak height velocity (13.8 years), while peak velocity of sitting height was 13.9 years (Nielsen *et al.*, 1986b). This is in accordance with the clinical observation that pubertal boys experience a stage of development with relatively long legs before attaining normal adult proportions.

The rise in testosterone excretion occurred approximately 2 years prior to peak height velocity and spermarche, but adult levels of testosterone were not obtained until 2 years after these events (Figure 4). This is compatible with the concept that high intratesticular testosterone concentrations, necessary for initiation of spermatogenesis, are reached a couple of years before adult levels of serum testosterone can be demonstrated. In addition, our results indicated that the highest

Figure 5. Urinary LH and testosterone (T) excretion, spermarche (vertical line), and growth velocities of height (HT), sitting height (H) and leg length (LL) in relation to age in 3 individual boys. Note the association of peak height velocity and spermarche. Reproduced with permission from Nielsen *et al* (1986b).

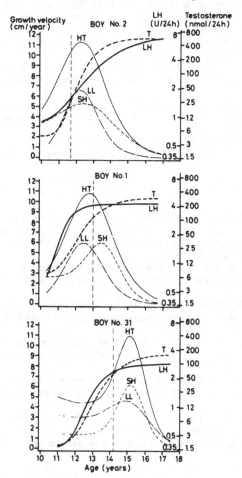

growth promoting influence of testosterone is attained at low concentrations of the hormone. At the time of maximal decelleration of height growth, testosterone excretion had almost reached those found in adult men.

Conclusion

The testis is endocrinologically active at very early stages of development. The activity seems to be associated with the development and maturation of the seminiferous epithelium even before puberty, whereas the significance for prepubertal growth has not yet been established. The marked pubertal increase in testosterone production of the testis seems to be associated with initiation of sperm production long before its maximal effect on height growth and development of secondary sex characteristics has been obtained.

References

Aynsley-Green, A., Zachmann, M. & Prader, A. (1976). Interrelation of the therapeutic effects of growth hormone and testosterone on growth in hypopituitarism. *Journal of Pediatrics*, **89**, 992–9.

Baldwin, B.T. (1928). The determination of sex maturation in boys by a laboratory method. *Journal of Comparative Psychology*, **8**, 39.

Cassorla, F.G., Golden, S.M., Johnsonbaugh, R.E., Heroman, W.M., Loriaux, D.L. & Sherins, R.J. (1981). Testicular volume during infancy. *Journal of Pediatrics*, **99**, 742–3.

Cortes, D., Müller, J. & Skakkebæk, N.E. (1987). Proliferation of Sertoli cells during human development assessed by stereological methods. *International Journal of Andrology*, in press.

Faiman, C., Winter, J.S.D. & Reyes, F.I. (1981). Endocrinology of the fetal testis. In *The Testis*, ed. H. Burger and D. de Kretser, p. 81. Raven Press, New York.

Forest, M.G., Cathiard, A.M. & Bertrand, J.A. (1973). Evidence of testicular activity in early infancy. *Journal of Clinical Endocrinology and Metabolism*, **37**, 148–51.

Forti, G., Santoro, S., Grisolia, G.A., Bassi, F., Boninsegni, R., Fiorelli, G. & Serio, M. (1981). Spermatic and peripheral plasma concentrations of testosterone and androstenedione in prepubertal boys. *Journal of Clinical Endocrinology and Metabolism*, **53**, 883–6.

Gendrel, D., Roger, M. & Job, J.-C. (1980). Plasma gonadotropin and testosterone values in infants with cryptorchidism. *Journal of Pediatrics*, **97**, 217–20.

Hirsch, M., Lunenfeld, B., Modan, M., Ovadia, J. & Shemesh, J. (1985). Spermarche – the age of onset of sperm emission. *Journal of Adolescent Health Care*, **6**, 35–9.

Karpe, B., Fredricsson, B., Svensson, J. & Ritzén, E.M. (1982). Testosterone concentration within the tunica vaginalis of boys and adult men. *International Journal of Andrology*, **5**, 549–56.

Müller, J. & Skakkebæk, N.E. (1984). Fluctuations in the number of germ cells during the late foetal and early postnatal periods in boys. *Acta Endocrinologica*, **105**, 271–4.

Nielsen, C.T., Skakkebæk, N.E., Darling, J.A.B., Hunter, W.M., Richardson, D.W., Jørgensen, M. & Keiding, N. (1986b). Longitudinal study of testosterone and luteinizing hormone (LH) in relation to spermarche, pubic hair, height, and sitting height in normal boys. *Acta Endocrinologica*, Suppl. **279**, 98–106.

Nielsen, C.T., Skakkebæk, N.E., Richardson, D.W., Darling, J.A.B., Hunter, W.M., Jørgensen, M., Nielsen, Aa., Ingerslev, O., Keiding, N. & Müller, J. (1986a). Onset of the release of spermatozoa (spermarche) in boys in relation to age, testicular growth, pubic hair, and height. *Journal of Clinical Endocrinology and Metabolism*, **62**, 632–5.

Preece, M.A. & Baines, M.J. (1978). A new family of mathematical models describing the human growth curve. *Annals of Human Biology*, **5**, 1–24.

Richardson, D.W. & Short, R.V. (1978). Time of onset of sperm production in boys. *Journal of Biosocial Science*, Supplement **5**, 15–25.

Siebert, J.R. (1982). Testicular weight in infancy. *Journal of Pediatrics*, **100**, 835–6.

Tanner, J.M., Whitehouse, R.H., Hughes, P.C.R. & Carter, B.S. (1976). Relative importance of growth hormone and sex steroids for the growth at puberty of trunk length, limb length, and muscle width in growth hormone-deficiency children. *Journal of Pediatrics*, **89**, 1000–8.

Winter, J.S.D. & Faiman, C. (1972). Pituitary–gonadal relations in male children and adolescents. *Pediatric Research*, **6**, 126–35.

INDEX